Systems Engineering and Program Management

Trends and Costs for Aircraft and Guided Weapons Programs

David E. Stem, Michael Boito, Obaid Younossi

Prepared for the United States Air Force

Approved for public release; distribution unlimited

The research reported here was sponsored by the United States Air Force under Contract F49642-01-C-0003. Further information may be obtained from the Strategic Planning Division, Directorate of Plans, Hq USAF.

Library of Congress Cataloging-in-Publication Data

Stem, David E.
Systems engineering and program management trends and costs for aircraft and guided weapons programs / David E. Stem, Michael Boito, Obaid Younossi.
p. cm.
Includes bibliographical references.
"MG-413."
ISBN 0-8330-3872-9 (pbk. : alk. paper)
1. United States. Air Force—Procurement—Costs. 2. Airplanes, Military—United States—Costs. 3. Guided missiles—United States—Costs. I. Boito, Michael, 1957– II. Younossi, Obaid. III. Title.

UG1123.S75 2006
358.4'162120973—dc22

2005030589

U.S. Air Force photo by Kevin Robertson

Published 2006 by the RAND Corporation
1776 Main Street, P.O. Box 2138, Santa Monica, CA 90407-2138
1200 South Hayes Street, Arlington, VA 22202-5050
4570 Fifth Avenue, Suite 600, Pittsburgh, PA 15213
RAND URL: http://www.rand.org/
To order RAND documents or to obtain additional information, contact
Distribution Services: Telephone: (310) 451-7002;
Fax: (310) 451-6915; Email: order@rand.org

Preface

Although systems engineering and program management (SE/PM) have long been part of aircraft and weapons systems development and production costs, there has not been a comprehensive, focused study that has addressed the issue of developing cost estimates for SE/PM. This report specifically focuses on techniques that can be used to estimate SE/PM costs. It also describes various functions within SE/PM and investigates possible cost drivers of SE/PM.

Through extensive interviews with government and industry personnel, a literature search of past studies regarding SE/PM cost analysis, and analysis of actual SE/PM data, the authors characterize trends in SE/PM costs and general estimating methodologies. This study should be of interest to government and industry cost analysts, the military aircraft and weapon acquisition community, and others concerned with current and future acquisition policies.

Because of its proprietary nature, the cost information for the relevant programs is provided in a supplementary RAND Corporation report (TR-311-AF), which is not available to the general public. Inquiries regarding the supplement should be directed to the Office of the Technical Director, Air Force Cost Analysis Agency at (703) 604-0387.

This report is one of a series of reports from a RAND Project AIR FORCE study entitled "The Cost of Future Military Aircraft: Historical Cost-Estimating Relationships and Cost-Reduction Initiatives." The purpose of the study is to improve the cost-estimating tools used to project the cost of future weapon systems and to investigate the effects of recent management initiatives and government policies on cost. The study is being conducted within the RAND Project AIR FORCE Resource Management Program. The research is sponsored by the Principal Deputy, Office of the Assistant Secretary of the Air Force (Acquisition), and by the Office of the Technical Director, Air Force Cost Analysis Agency.

Other RAND Project AIR FORCE reports that address military aircraft cost estimating issues are the following:

- *Military Airframe Acquisition Costs: The Effects of Lean Manufacturing*, Cynthia R. Cook and John C. Graser (MR-1325-AF, 2001). This report examines the package of new tools and techniques known as "lean production" to determine if it would enable aircraft manufacturers to produce new weapons systems at costs below those predicted by historical cost estimating models.
- *An Overview of Acquisition Reform Cost Savings Estimates*, Mark A. Lorell, John C. Graser (MR-1329-AF, 2001). In this report, the authors examine relevant literature and conducted interviews to determine whether estimates on the efficacy of acquisition reform measures are sufficiently robust to be of predictive value.
- *Military Airframe Costs: The Effects of Advanced Materials and Manufacturing Processes*, Obaid Younossi, Michael Kennedy, John C. Graser (MR-1370-AF, 2001). This report examines the effect of the use of advanced materials, such as composites and titanium, on military aircraft costs. The report provides cost estimators with useful factors for adjusting and creating estimates based on parametric cost-estimating methods.
- *Military Jet Engine Acquisition: Technology Basics and Cost-Estimating Methodology*, Obaid Younossi, Mark V. Arena, Richard M. Moore, Mark A. Lorell, Joanna Mason, John C. Graser (MR-1596-AF, 2002). This report updates earlier studies in the area of propulsion cost analysis, discusses recent engine technologies, and provides methods and techniques that can be used to estimate the costs of future engine programs.
- *Test and Evaluation Trends and Costs for Aircraft and Guided Weapons*, Bernard Fox, Michael Boito, John C. Graser, Obaid

Younossi (MG-109-AF, 2004). This report examines the effects of changes in the test and evaluation (T&E) process used to evaluate military aircraft and air-launched guided weapons during their development programs. It also provides relationships for developing estimates of T&E costs for future programs.

- *Software Cost Estimation and Sizing Methods: Issues and Guidelines*, Shari Lawrence Pfleeger, Felicia Wu, Rosalind Lewis (MG-269-AF, 2005). This report recommends an approach to improve the utility of software cost estimates by exposing uncertainty and reducing risks associated with developing software estimates.
- *Lessons Learned from the F/A-22 and F/A-18E/F Development Programs*, Obaid Younossi, David E. Stem, Mark A. Lorell, Frances M. Lussier (MG-276-AF, 2005). This reports evaluates the history of the F/A-22 and F/A-18 E/F programs to understand how costs and schedules changed during their development. The study derives lessons that the Air Force and other services can use to improve acquisition.

RAND Project AIR FORCE

RAND Project AIR FORCE (PAF), a division of the RAND Corporation, is the U.S. Air Force's federally funded research and development center for studies and analyses. PAF provides the Air Force with independent analyses of policy alternatives affecting the development, employment, combat readiness, and support of current and future aerospace forces. Research is performed in four programs: Aerospace Force Development; Manpower, Personnel, and Training; Resource Management; and Strategy and Doctrine.

Additional information about PAF is available on our Web site at http://www.rand.org/paf.

Contents

Figures

Tables

Summary

Background

Sound cost estimates are essential to developing good budgets and policy decisions. Some recent RAND studies have looked at estimating techniques for the nonrecurring and recurring flyaway costs of military airframes and engines. This study extends the analysis into what are termed "below-the line" costs.[1] Below-the-line costs include costs for such items as system test and evaluation, data, special test equipment and tooling, training, operational site activation, industrial facilities, initial spares and repair parts, and systems engineering and program management. These costs are not directly associated with the development or the production of the hardware end item. Nevertheless, they are important cost elements that are necessary for delivery of the complete end item to the government.

RAND began the investigation of below-the-line costs with a study of systems test and evaluation costs (Fox et al., 2004). As a follow-on to that earlier study, this study investigates cost-estimating techniques that can be used to estimate Systems Engineering and

[1] Cost estimates for the Department of Defense are usually structured around the product-centric work breakdown structure described in Military Handbook 881 (Mil-HDBK-881). The handbook provides a framework for categorizing program costs starting with the hardware and software costs directly associated with the end item and adding the below the line costs. Below the line costs derive their name from the fact that they are typically displayed in budget documents and cost estimates as separate cost elements below the hardware cost elements.

Program Management (SE/PM) costs for military aircraft and weapons systems in development and production.

Analysis Approach

Our approach to analyzing SE/PM costs was to first understand the nature of the content of the work that is performed in this area. We define what is involved in the systems engineering and program management disciplines from a general sense of what SE/PM is and describe the iterative process and tools (such as reviews and documents that are developed for a program) that are used in the field. The definition and processes provide a basis for understanding what makes up the scope of the SE/PM effort.

Our next step was to canvas government and industry personnel to learn about the current state of techniques used for estimating SE/PM and to gather data that could be used to investigate various aspects of SE/PM costs. We used a questionnaire, presented in Appendix C, and interviews with both government and contractor personnel to find out how they define SE/PM costs, what techniques they currently use to estimate SE/PM costs, and what they would consider potential cost drivers that could be used for predicting costs. To develop SE/PM cost estimating methods, we collected historical data from several aircraft and missile development and production programs. The data included historical costs, the schedule of major events in the program, and technical information from several aircraft and missile programs. Cost data were collected from a variety of government cost reports and internal contractor accounting reports on programs from the 1960s to today. These data were used to investigate trends in SE/PM costs over time and to generate cost estimating methodologies that cost analysts could use when little program information is available early in the lifecycle of a program.

Our last step was to investigate the effects of new acquisition initiatives on SE/PM costs. The three new acquisition initiatives we investigated were the removal of military specifications and standards, the use of integrated product and process teams, and the relatively new preferred acquisition approach of evolutionary acquisition. Each of these initiatives could affect SE/PM costs. We tried to determine whether the SE/PM costs for these types of programs were different enough from the SE/PM costs for traditional acquisition programs that some adjustment in cost estimating should be made.

Definitions and Methods

One of the complications in developing SE/PM estimates is determining what is included in SE/PM costs. We found that the definition used by the government, as spelled out in (MIL-HDBK-881) and excerpted in Appendix B of this report, covers tasks associated with the "overall planning, directing, and controlling of the definition, development, and production of a system . . . [but] *excludes* systems engineering and program management effort that can be associated specifically with the equipment (hardware/software) element." The exclusionary portion of the definition is difficult to implement because the systems engineering associated with a program is integral to the development of the hardware and software of the system.

When recording SE/PM costs that are incurred, contractors' accounting systems may not consistently address this exclusion in the SE/PM definition. After we interviewed multiple contractors and investigated their detailed internal accounting data, we found their costs under the SE/PM category were not always consistent (see page 54). Some of this difference across contractors was anticipated due to variations in accounting methods. We further found that even within a single company there were differences from one program to another as to what was classified as SE/PM costs (see page 57). Although these differences exist, based on an examination of detailed cost data, we believe that the main cost sub-elements that represent a large portion of the SE/PM costs are classified consistently across contractors and programs (see page 57).

Our discussions with government personnel and contractors revealed a variety of techniques for estimating SE/PM costs. In general, for aircraft programs that are early in their acquisition lifecycle, the government estimates SE/PM using a parametric approach applied at an aggregate level that includes not only the costs of SE/PM, but also includes the costs associated with hardware design. The parametric approaches rely on independent parameters that relate to the overall design of a system (i.e., weight, speed, first flight). This approach is consistent with the task of the government cost estimator—to generate a budget estimate that includes all expected costs, regardless of how they are classified. However, this high level of estimating does not allow for understanding the cost drivers specifically associated with SE/PM and how SE/PM costs are expended through a multiyear development program. This approach also makes it difficult to isolate SE/PM costs for any potential adjustments due to acquisition changes that may have a cost impact. The industry contacts we interviewed used a variety of techniques for developing SE/PM estimates, ranging from "top-down" models to "bottom-up" approaches. The type of model they use generally depends on the desired level of fidelity and level of detail of the estimate and on the maturity of the program. Top-down models typically use parametric approaches similar to those the government uses when little detailed information is known about a program. Bottom-up approaches are used as a program becomes more mature and better information is available that allows more-detailed comparisons with prior experiences.

Because our objective for this study was to develop methodologies that could be used to estimate SE/PM costs directly, we used statistical analysis to develop parametric cost-estimating relationships (CERs) for aircraft and guided weapons programs in development and production. We wanted our resulting estimating methods to utilize parameters directly related to SE/PM costs. Based on our interviews with contractor personnel and on previous cost studies, we generated a list of potential independent variables that could be logically related to SE/PM costs. Using step-wise and ordinary least squares regression analysis, we selected the best CERs most useful to predicting costs.

Finally, to determine if any adjustments to historically based CERs are required to account for new acquisition approaches, we compared the SE/PM cost data from selected sample programs that implemented the new acquisition approaches with the SE/PM cost data for the overall sample of similar programs. We wanted to see what, if any, differences arose in the SE/PM cost for these programs under acquisition reform as compared with other programs to determine if any changes to our estimating methods were necessary to take these new initiatives into account.

Results and Findings

We first examined historical SE/PM costs over time to determine what general cost trends seem to be occurring. As the basis for our quantification of SE/PM development costs, we collected data from a wide assortment of historical efforts including prototype development programs, full development programs, and modification programs. For production analysis, we also used data from several production lots from multiple programs. The data we gathered on aircraft and weapons programs from the 1960s up through recent years showed that SE/PM represents a significant portion of program cost and seems to be on the rise for aircraft development programs (see page 29). For aircraft development programs, SE/PM represents about 12 percent of the total contractor cost. For weapons development programs, the SE/PM percentage of the total cost is even larger—28 percent on average. We found the SE/PM cost split between systems engineering and program management is roughly 50/50 for aircraft programs and 60/40 for weapons programs (see page 34). SE/PM production data for aircraft showed a large amount of variation, while production cost for weapons seemed to more closely follow a traditional cost-improvement curve (also referred to as a learning curve) (see page 36).

Based on our interviews with contractor personnel and a review of prior studies of aircraft and weapons costs, we explored a set of independent variables that we believe could be related to SE/PM cost. Most independent variables fell into three categories: program scope variables, programmatic variables, and physical descriptor variables. Program scope is measured by the cost of the program less SE/PM costs in either development or production. Programmatic variables capture the duration of the effort (in the case of development programs) and quantity of items produced (in the case of production programs). Physical descriptor variables are generally weight based, except in the case of weapons for which diameter was also considered. In addition to these variables, for aircraft in development, we attempted to relate the amount of integration required (as measured by air vehicle cost divided by airframe cost) to the overall SE/PM cost. For weapons in development, we also considered programmatic variables to account for programs that were not traditional engineering and manufacturing development (EMD) programs (i.e., prototype programs or modification programs) and to account for changes over time (based on the contract award year). We were also sensitive to using independent variables that could readily be quantified by a cost analyst early in a program.

We found that for both aircraft and weapons in development, SE/PM costs were most directly related to the overall size of the program (as measured by development cost less SE/PM) (see page 79). In addition, we found that design duration (as measured by months from contract award to first flight) played a role in the SE/PM cost for aircraft development programs (see page 80). In looking at the funding profile of SE/PM costs, we found that about one-third of the total SE/PM cost is expended from contract award to critical design review, the second third of the SE/PM cost is spent from CDR to the first flight date, and the final third is spent from first flight to the end of the program (see page 82). Appendix F details techniques that can be used to time-phase SE/PM development cost estimates.

For both aircraft and weapons, we again determined that scope (as measured by the recurring unit cost of the aircraft or missile) was a significant factor in estimating SE/PM cost in the production phase of the program. In addition, we determined that the cumulative quantity and production rate were related to the unit cost of SE/PM in production (see pages 88 and 105). The ratio of the yearly lot size to the maximum lot size was found to be an independent variable that improved the predictive capability of our estimating equations (see page 93). The cost-improvement slopes, used for projecting yearly SE/PM costs, showed a large variation for aircraft programs, while the slopes for weapons were more tightly grouped (see pages 98 and 111).

Unfortunately, the large degree of variation in the data we used to develop these parametric estimating methods resulted in a large standard error for our estimating equations. We tried to further investigate what might be causing the variation, but were unable to identify any consistent cause. For example, in the case of aircraft production costs, we looked to see whether the high degree of SE/PM cost variability was related to the change in the aircraft model or to the introduction of foreign military sales. These two changes did not align with the fluctuations in the SE/PM cost data (see page 84). For these reasons, we conclude that the CERs we generated are most useful to a cost estimator in the early stages of a program's life cycle, when little is known about the program. When more detailed information is available, other techniques could be applied for developing more-accurate SE/PM estimates. For example, use of a direct-analogy approach in which a well-understood program is compared with a new program can lead to less variation in the final outcome and a better understanding of the specific cost drivers (see page 125).

Finally, we investigated the potential effect that new acquisition approaches, such as decreased use of military specifications and military standards, use of integrated product teams (IPTs), and the use of evolutionary acquisition, would have on SE/PM costs. Because there is not a long history of these types of programs, we compared the SE/PM costs of the few programs that have implemented these changes to the overall population of similar programs. We found that programs that minimized military specifications did not show a significant difference from the overall sample of programs, being within one standard deviation in SE/PM cost from the overall sample average (see page 114). For programs that used IPTs, SE/PM costs were either similar to or slightly higher than SE/PM costs for the overall sample of programs (see page 116). To determine the quantitative effect that evolutionary acquisition (EA) had on SE/PM costs, we analyzed SE/PM costs for a program that concurrently developed multiple variants as a surrogate for an EA program, and we found that it exhibited above-average SE/PM costs (see page 121). In addition, we investigated cost-estimating methodologies employed by one of the first programs to use EA. The cost-estimating technique used by one formally designated evolutionary acquisition program suggests that two areas of SE/PM need to be estimated: the SE/PM cost associated with the specific capability increment or "spiral" and the "overlay" SE/PM cost that is concerned with development and production of the overall program (see page 120).

In conclusion, SE/PM costs are a large portion of the acquisition cost of military aircraft and guided weapons systems. In the case of aircraft, SE/PM costs appear to be rising over time. There are multiple approaches to estimating the cost of SE/PM, and each has advantages and disadvantages. We developed a set of cost-estimating relationships that can be used to specifically estimate the SE/PM cost element for development and production for both aircraft and weapons programs. However, the production CERs we generated were not as good as the development CERs at explaining the variation in the historical data. Finally, we found that implementation of new acquisition approaches had mixed results in changing SE/PM costs.

Acknowledgments

Many individuals were instrumental in the completion of this study. A number of individuals, from both government and industry, took the time to provide insights and information used in this study.

This study would not have been possible without the sponsorship of Lt Gen John Corley, the U.S. Air Force Principal Deputy Assistant Secretary for Acquisition. Also, we greatly appreciate the oversight provided by the Air Force Cost Analysis Agency and its assistance in gathering the data used for the analysis in this report. In particular, Scott Adamson and John Fitch were instrumental in giving us access to government cost data.

From the government, we were able to obtain insights and data from the following organizations and individuals:

- Office of Secretary of Defense (Cost Analysis Improvement Group): Edward Kelly and Fred Janicki
- Office of Secretary of Defense (Acquisition, Technology & Logistics): James Thompson
- Joint Strike Fighter System Program Office: Michael Clark
- Naval Air Systems Command, Cost Department: Richard Scott, Heidi Farmer, and William Stranges
- Aeronautical Systems Command: Mike Seibel and Sandra McCardle
- Naval Center for Cost Analysis: Thomas Burton
- Global Hawk System Program Office: Walt Pingle.

The following industry representatives whom we interviewed provided useful insights that were crucial to our developing a deeper understanding of systems engineering and program management tasks and the cost drivers associated those tasks:

- Boeing Corporation: Carol Hibbard, Dru Held, Kimberly Schenken, and Timothy Stremming
- Raytheon Corporation: David Sauve
- Lockheed Martin: Ralph Smith and Cleo Lyles.

The authors would also like to thank RAND Corporation colleagues for making significant contributions to this report. We would like to thank Jim Thompson from the Office of Undersecretary of Defense and RAND colleague Bernard Fox for their thoughtful and careful review. Jack Graser provided much-needed guidance and direction for the analysis effort. Allan Crego assisted with the statistical analysis of the data. Nancy DelFavero did an outstanding job of editing and improving the flow of the discussion in the final report, and Jennifer Li and Jane Siegel made an outstanding contribution to structuring a previous version of this report. Michele Anandappa, Mike DuVal, and Nathan Tranquilli provided valuable administrative support.

Acronyms and Abbreviations

AAAM	Advanced Air-to-Air Missile
AAD	Airframe Already Developed
ACAT	Acquisition Category
ACTD	Advanced Concept Technology Demonstration
AFCAA	Air Force Cost Analysis Agency
AIA	Aircraft Industry Association
ALCM	Air-Launched Cruise Missile
AMRAAM	Advanced Medium-Range Air-to-Air Missile
ANSI	American National Standards Institute
ASD(C3I)	Assistant Secretary of Defense for Command, Control, Communications, and Intelligence
ASD(NII)	Assistant Secretary of Defense (Networks and Information Integration)
ASR	Alternate System Review
ATA	Advanced Tactical Aircraft
ATF	Advanced Tactical Fighter
AV/AF	Air Vehicle/Airframe (cost)
AV AUC	Air Vehicle Average Unit Cost
AV T100	Air Vehicle 100th Unit Cost
AWACS	Airborne Warning and Control System
CA	Contract Award

CADT	Contract Award to End of Development Test
CAFF	Contract Award to First Flight
CAFGL	Contract Award to First Guided Launch
CAFPD	Contract Award to First Production Delivery
CAIV	Cost as an Independent Variable
CCDR	Contractor Cost Data Report
CDD	Capabilities Development Document
CDR	Critical Design Review
CDRL	Contract Data Requirements List
CER	Cost Estimating Relationship
CMM	Capability Maturity Model
CPD	Capability Production Document
C3I	Command, Control, Communications, and Intelligence
CPR	Cost Performance Report
CWBS	Contractor Work Breakdown Structure
DAU	Defense Acquisition University
DEN	Density
DIAM	Diameter
DoD	Department of Defense
DoDD	Department of Defense Directive
DSMC	Defense Systems Management College
DT	Development Test
DTC	Design to Cost
DT&E	Development Test and Evaluation
D&V	Demonstration and Validation
DVMOD	Demonstration and Validation or Modification
EA	Evolutionary Acquisition
ECM	Electromagnetic Compatibility

ECP	Engineering Change Proposal
EIA	Electronic Industries Alliance
EMD	Engineering and Manufacturing Development
ESWBS	Extended Ship Work Breakdown Structure
EVM	Earned Value Management
FCA	Functional Configuration Audit
FCHR	Functional Cost Hour Reports
FF	First Flight
FMS	Foreign Military Sales
FQR	Formal Qualification Review
FRP	Full-Rate Production
FSD	Full-Scale Development
FSED	Full-Scale Engineering Development
G&A	General and Administrative
FY	Fiscal Year
GFE	Government-Furnished Equipment
GPS	Global Positioning System
HARM	High Speed Anti-Radiation Missile
IBR	Integrated Baseline Review
ICBM	Intercontinental Ballistic Missile
ICD	Initial Capabilities Document
IDA	Institute for Defense Analyses
IEC	International Electrotechnical Commission
IEEE	Institute of Electrical and Electronics Engineers
IIPT	Integrating Integrated Product Team
ILS	Integrated Logistics Support
ILSP	Integrated Logistics Support Plan
IMU	Inertial Measurement Unit
INCOSE	International Council on Systems Engineering

IPPD	Integrated Product and Process Development
IPT	Integrated Product Team
IR	Infrared
ISO	International Organization for Standardization
ISR	In-Service Review
ITR	Initial Technical Review
IV&V	Independent Verification and Validation
JASSM	Joint Air-to-Surface Standoff Missile
JDAM	Joint Direct Attack Munition
JSF	Joint Strike Fighter
JSOW	Joint Standoff Weapon
KPP	Key Performance Parameter
LCC	Life-Cycle Cost
LLOT	Last Lot
LOR	Level of Repair
LOT MP	Lot Midpoint
LOT NUM	Lot Number
LRIP	Low-Rate Initial Production
LRU	Line Replaceable Unit
LSA	Logistics Support Analysis
MDA	Milestone Decision Authority
MDAP	Major Defense Acquisition Program
MIL-HDBK	Military Handbook
MILSPEC	Military Specification
MILSTD	Military Standard
MMP	Manufacturing Management Plan
NCCA	Naval Center for Cost Analysis
NNTE	Nonrecurring, Nontest Engineering
NRDEV	Nonrecurring Development Cost

OIPT	Overarching Integrated Product Team
O&S	Operating and Support
OSD(PA&E)	Office of Secretary of Defense (Program Analysis and Evaluation)
OT&E	Operational Test and Evaluation
OTRR	Operational Test Readiness Review
PAF	Project AIR FORCE
PCA	Physical Configuration Audit
P&D	Production and Deployment
PDR	Preliminary Design Review
P3I	Preplanned Product Improvement
PIN	Program Integration Number
PPI	Post-Production Improvement
PM	Program Management
PROTO	Prototype
PRR	Production Readiness Review
RATE	Rate per Year
RDT&E	Research, Development, Test, and Evaluation
RFP	Request for Proposal
SAR	Selected Acquisition Report
SDP	Software Development Plan
SD&D	System Development and Demonstration
SE	Systems Engineering
SEI	Software Engineering Institute
SEMP	Systems Engineering Management Plan
SE/PM	Systems Engineering/Program Management
SEPM LOT	Lot Cost of SE/PM (in production)
SEPM UC	Unit Cost of SE/PM (in production)
SFR	System Functional Review

SLCM	Sea-Launched Cruise Missile
SLOC	Source Lines of Code
SOO	Statement of Objectives
SOW	Statement of Work
SPEC	Specification
SRAM	Short-Range Attack Missile
SRR	System Requirements Review
SVR	System Verification Review
ST&E	Systems Test and Evaluation
T100	100th Unit
T1000	1000th Unit
TDEVLSEPM	Total Development Cost Less SE/PM Cost
TEMP	Test and Evaluation Master Plan
TRA	Technology Readiness Assessment
TRL	Technical Readiness Level
TRR	Test Readiness Review
TSSAM	Tri-Service Standoff Attack Munition
USD(AT&L)	Under Secretary of Defense (Acquisition, Technology, and Logistics)
VECP	Value Engineering Change Proposal
WBS	Work Breakdown Structure
WE	Weight Empty
WIPT	Working-Level Integrated Product Team
WPN	Weapon
WT	Weight

CHAPTER ONE

Introduction

Study Background and Purpose

For cost estimators to develop sound estimates for program budgets, reliable and accurate cost-estimating techniques are needed. As new programs are fielded and as acquisition management techniques change, there is a constant need to improve the tools available to cost estimators. This report explores cost data on historical aircraft and guided weapons programs and presents techniques for developing sound estimates of systems engineering and program management (SE/PM) costs.

In the Department of Defense (DoD), cost estimates and budgets are structured to follow a product-centric, work breakdown structure that itemizes program tasks and costs in a hierarchical fashion. Guidelines for developing a work breakdown structure (WBS) are described in Military Handbook 881 (MIL-HDBK-881). The suggested generalized WBS for aircraft is shown in Table 1.1.

The WBS provides a common structure for understanding and allocating tasks, expending resources, and reporting to the government. Each lower level of the WBS has a "child-to-parent" relationship such that Level 1 encompasses the entire aircraft system. Level 2 costs include air vehicle costs that are associated with hardware and software that make up the complete flying aircraft. The Level 3 elements, which are under air vehicle, include the airframe, propulsion, and all other installed equipment. The other Level 2 cost elements that begin with SE/PM are typically termed "below-the-line" costs.

Table 1.1
Generic Aircraft System Work Breakdown Structure

Level 1	Level 2	Level 3
Aircraft System		
	Air Vehicle	
		Airframe
		Propulsion
		Air Vehicle Applications Software
		Air Vehicle System Software
		Communications/Identification
		Navigation/Guidance
		Central Computer
		Fire Control
		Data Display and Controls
	Systems Engineering/Program Management	
	Systems Test and Evaluation	
	Training	
	Data	
	Peculiar Support Equipment	
	Common Support Equipment	
	Operational/Site Activation	
	Industrial Facilities	
	Initial Spares and Repair Parts	

NOTE: Below-the-line costs are shown in italics.

The below-the-line cost elements (shown in Table 1.1) are common across multiple types of systems that DoD develops and procures. *Systems engineering and program management* costs include the costs of business management as well as the costs of engineering and technical control of a particular program. *Systems test and evaluation* costs are the costs associated with using specific hardware and software to validate that the engineered design meets the desired performance of the system. *Training* costs include the costs of services and equipment to instruct personnel in the operation and maintenance of the system. The *data* cost element includes the costs of delivering to the government data associated with the contract. *Peculiar support equipment* (as it is called in MIL-HDBK-881) covers the cost of developing and producing system-specific equipment to support and maintain the system. *Common support equipment* is associated with items currently in the DoD inventory that are required to support and maintain other systems. *Operational/site activation* costs are the costs associated with the facilities to house and operate the system. *Industrial facilities* costs are the depot maintenance start-up costs. *Initial spares and repair parts* costs are the costs for initial spares for a newly fielded system.

SE/PM represents one of the more costly of the below-the-line elements for military aircraft and guided weapon systems. This report explores the content of the work performed under SE/PM and looks at the trends that have been occurring in SE/PM costs for both development and production for aircraft and weapons. We also discuss current methods used by government and industry to estimate the cost of SE/PM, and we provide some useful cost-estimating relationships (CERs) that can be used for programs early in their development. The parameters used in the CERs show how SE/PM costs can be estimated by knowing some basic information about the program.

We developed parametric cost-estimating approaches that could be used to directly estimate SE/PM costs as a separate WBS element. (Traditional estimating approaches estimate SE/PM as part of the larger design effort.) We found cost drivers specifically related to SE/PM costs. These cost drivers are quantifiable and determined early in the acquisition of a program. We used regression analysis to determine how these cost parameters can forecast SE/PM costs.

Another goal of this study is to determine if the DoD's recent acquisition-process initiatives—collectively referred to as "acquisition reform"—affect SE/PM costs. Specifically, we investigate the effect of three acquisition reform initiatives—the reduction of the number of military specifications and standards, the use of integrated product teams, and the use of evolutionary acquisition—on SE/PM costs. Because these processes have changed the traditional acquisition process, we wanted to determine what, if any, impact these changes are likely to have on SE/PM costs.

Comparison with Previous Work in This Area

Several studies have been performed that discuss methodologies that can be used to estimate the development and production cost of aircraft and guided weapons programs. Most of the studies focus on estimating higher-level elements of cost (i.e., total engineering) rather than just focusing on the SE/PM costs associated with a program. Typically, SE/PM costs are grouped with engineering costs and are not estimated separately. A description of some of the previous work in this area is provided below.

RAND has done research in the area of cost analysis for military aircraft and weapons since the 1950s. Many of these reports use historical cost information to develop parametric estimating equations that link cost to various independent variables that usually measure physical or performance characteristics of the systems. In 1987, RAND developed a series of CERs for estimating airframe costs for military aircraft programs (Hess and Romanoff, 1987) that used aircraft weight empty (WE) and speed as independent variables.

Over time, the increased complexity of aircraft has created the desire to investigate other independent variables to predict airframe costs. In 1991, RAND published a report (Resetar, Rogers, and Hess, 1991) that looked at the relative cost of using advanced materials (such as composites) as compared with the cost of using traditional metal materials in aircraft development and production. This study was updated in 2001 (Younossi, Kennedy, and Graser, 2001) and provided a CER for estimating total nonrecurring engineering hours to develop an airframe. Included in the CER is the cost of the associated SE/PM effort; however, the study's report does not present a method for extracting SE/PM costs.

In 1988, the Institute for Defense Analyses (IDA) published a study (Harmon et al., 1988) commissioned by the Office of the Secretary of Defense (OSD), Program Analysis and Evaluation (PA&E) that investigated the costs of developing military tactical aircraft. This report was prepared as the Air Force was about to begin the Advanced Tactical Fighter (ATF) program and the Navy was about to start the Advanced Tactical Aircraft (ATA) program. The IDA report contains descriptive overviews and costs related to the development of 23 different programs from the 1960s to the 1980s.

As did the prior RAND and IDA reports, this report provides parametric cost-estimating approaches that can be used to generate cost estimates at the beginning of a program's development before much detail is known about the program's design. It is different from the prior RAND and IDA studies in that it provides a methodology for estimating the prime contractor's SE/PM costs directly.

Study Methods and Approach

DoD has in recent years placed a greater emphasis on the rigorous application of a systems engineering approach to all programs within the department. The acting Undersecretary of Defense for Acquisition, Technology and Logistics (USD[AT&L]), Michael Wynne, stated in a memo that "all programs . . . regardless of acquisition category, shall apply a robust SE approach" (USD[AT&L], 2004). Given the highlighted importance of this approach, and the fact that SE/PM tends to represent a large portion of the cost for aircraft and guided weapons systems, it is worthwhile to address the content and costs associated with SE/PM.

As stated above, the purpose of this study is to investigate the costs that are specifically categorized as SE/PM costs for aircraft and weapons programs in development and production. In contrast to prior studies that grouped SE/PM costs with other engineering development costs, we developed CERs for estimating SE/PM costs directly using explanatory variables that are more closely associated with the SE/PM effort.

This method would then lend itself to further investigation of what changes, if any, should be considered for future SE/PM estimates given recent changes in the acquisition environment. One recent change is the evolution of prime contractors toward a greater lead systems-integrator role, with much of the detailed design and manufacturing outsourced to lower-level suppliers. Another change is that contractors are more involved in deriving specifications from top-level performance requirements rather than using established military specifications and standards. DoD programs have incorporated Integrated Product Teams (IPTs) that require increased levels of communication and coordination. Evolutionary acquisition and the use of cost as an independent variable (CAIV) have changed the acquisition approach by generating more design iterations through the development process rather than using a somewhat fixed design throughout development.

Our approach was to start by defining SE/PM and understanding the SE/PM process that is used by defense contractors in the acquisition of military equipment. We provide background on the tools (such as formal documentation and program reviews) that are commonly used in DoD programs to provide final products that meet the user's requirements. Most of this background information came from a literature search of related DoD acquisition handbooks and texts.

To better understand trends in SE/PM costs and methods currently used to estimate SE/PM, we interviewed both government and industry personnel who perform such estimates. We developed a questionnaire (provided in Appendix C) that poses questions to personnel working in the defense industry about the definitions they follow and the methods they use in estimating SE/PM costs. When we interviewed contractor personnel, we further asked them what relevant independent variables could be used to develop parametric estimating relationships. Finally, we asked what effects the new acquisition initiatives were having (if any) on SE/PM activities and costs.

The cost data we collected for this study generally came from contractor cost data reports (CCDRs) that covered the initial system development or system production. This data were supplemented by data from cost performance reports (CPRs) and from contractors' internal cost-reporting systems to gain a more thorough understanding of the cost details at lower levels of cost. The data set includes a mix of actual costs from programs that completed development, are currently in development, or that were canceled during development. For aircraft systems, we collected cost information on 26 development programs and 13 production programs. For guided weapon systems, we collected cost data from 37 development programs and

14 production programs. Tables 1.2 through 1.5 list information on the programs we investigated and the type of cost information that we used for this study (CCDRs, CPRs, and contractor/program data). For guided-weapons production programs, all the data are from CCDRs provided to the government.

Table 1.2
Aircraft Development Programs, Program Phases, and Sources of Cost Data

Program Name	Phase	CCDR	CPR	Contractor/ Program Data
B-1A	FSD	X		
B-1B	FSD	X		
B-1B CMUP Blk D	Modification development	X		
B-2	FSD		X	
C-5A	FSED	X		
C-17	FSED	X		
A-10	FSD	X		
AV-8B	FSD	X		
E-2C	Modification development	X		
E-3	FSD	X		
E-8	FSD	X		
767 AWACS	Modification development			X
Air Force One	Modification development			X
F-14	FSD (Lot I & II)	X		
F-15	FSD (Segment I & II)	X		
F-16	FSD	X		X
F/A-18A/B	FSD	X		X
F/A-18E/F	EMD	X		X
F-22	EMD	X		X
YF-16	Prototype development			X
YF-17	Prototype development			X
YF-22	Prototype development	X		
S-3	Development	X		
T-45	FSD	X		
V-22	FSD	X		
V-22	EMD	X		

NOTES: FSD = full-scale development; FSED = full-scale engineering development; EMD = engineering and manufacturing development.

Table 1.3
Aircraft Production Programs, Years of Program, and Sources of Cost Data

Program Name	Years	CCDR	CPR	Contractor/Program Data
F-14A	1970–1980	X		X
F-14A Plus	1986–1988	X		X
F-14D	1988–1990	X		X
F-15A/B	1973–1977	X		
F-15C/D	1978–1986	X		
F-15E	1987–1991	X		
F-16A/B	1977–1982	X	X	X
F-16C/D	1983–1993	X	X	X
A-10	1975–1982	X		
A-6A	1966–1969	X		
A-6E	1970–1988	X		
EA-6B	1967–1989	X		
AV-8B	1982–1991	X		
C-17	1988–1996	X	X	
C-5A	1966–1968	X		
C-5B	1984–1986	X		
E-2C	1971–1992	X		
E-3	1975–1978	X		
F/A-18A/B	1979–1985	X		

Table 1.4
Guided Weapons Development Programs, Program Phases, and Sources of Cost Data

Program Name	Phase	CCDR	CPR	Contractor/ Program Data
Air-to-Air				
AIM-54A	FSED	X		
AIM-54C	FSED	X		
AIM/RIM-7M	FSED	X		
AIM/RIM-7P	FSED	X		
AIM-155 (AAAM – General Dynamics and Westinghouse)	FSED	X		
AIM-155 (AAAM - Hughes and Raytheon)	FSED	X		
AIM-120	FSED	X		

Table1.4—Continued

Program Name	Phase	CCDR	CPR	Contractor/ Program Data
AIM-120	FSED 2nd Source	X		
AIM-120 P3I Phase I	EMD P3I			X
AIM-120 P3I Phase II	EMD P3I			X
AIM-120 P3I Phase III	EMD P3I			X
AIM-9R	FSED	X		
AIM-9X	D&V	X		
AIM-9X	EMD	X		
Air-to-Ground				
AGM-69A	FSED	X		
AGM-131A/B	FSD	X		
AGM-65A	FSED	X		
AGM-65D	FSED	X		
AGM-65C	FSED	X		
AGM-84A	Design and Weapons System Development	X		
AGM-84H	EMD	X		
AGM-88A, Sub Phase I	FSED	X		
AGM-88A, Sub Phase II	FSED		X	
AGM-88A, Sub Phase III	FSED		X	
AGM-86A	FSED	X		
AGM-86B	FSD	X		
AGM/RGM-136	FSED	X		
AGM-129A/B	FSED (Guidance)	X		
AGM-130	FSED	X		
CBU-97/B	FSED		X	
AGM-154A	EMD	X		
AGM-154B	EMD	X		
AGM-154C	EMD	X		
GBU-31	EMD (Phase II)	X		
AGM-158	EMD	X		
BGM-109	FSD	X		
R/UGM-109E	EMD		X	

NOTES: P3I = pre-planned product improvement; AAAM = Advanced Air-to-Air Missile; D&V = demonstration and validation.

Table 1.5
Guided Weapons Production Programs, Years of Programs, and Sources of Cost Data

Program Name	Years	CCDR
ALCM	1979–1983	X
AIM-120 Hughes	1987–1994	X
AIM-120 Raytheon	1987–1993	X
AGM-88AB HARM	1981–1992	X
AGM-84 Harpoon	1975–1987	X
AGM-65A Maverick	1987–1991	X
AGM-65 IR Maverick Hughes	1977–1982	X
AGM-65 IR Maverick Raytheon	1983–1993	X
AIM-54A Phoenix	1975–1982	X
AIM-9M Raytheon	1966–1969	X
AIM-9L Raytheon	1970–1988	X
SLCM Tomahawk	1967–1989	X
AIM-7M Sparrow General Dynamics	1982–1991	X
AIM-7M Sparrow Raytheon	1988–1996	X
AIM-7F Sparrow Raytheon	1966–1968	X

NOTES: ALCM = air-launched cruise missile; HARM = high-speed anti-radiation missile; IR = infrared; SLCM = sea-launched cruise missile.

We then combined the information collected through the interviews along with the cost data collected from government reports to generate a set of cost relationships that can be used to directly estimate SE/PM costs early on in a program when, generally, there is very little known about the specific characteristics of the program design. Based on the interviews and our own analysis, several independent variables were chosen as potential explanatory parameters. Our statistical analysis included performing a step-wise regression analysis to determine which independent variables would be the best to include in a cost model. After choosing the independent variables, we used ordinary least squares regression analysis to arrive at cost estimating equations for SE/PM costs as a function of these independent variables.

For the final part of our analysis, we used the cost data to investigate what effects new acquisition processes might have on SE/PM costs. We used univariate analysis to determine if there were any differences between the SE/PM costs for these individual programs using the new acquisition processes and the SE/PM costs for the total sample of programs.

Limitations of the Study

After interviewing personnel from defense companies, it became clear that each of the companies, and in some cases each program within a company, defines SE and PM somewhat differently. Part of the reason for these differences is that much of the defense industry has consolidated into a few prime-level contractors. The new merged companies are products of several legacy companies that each used different accounting systems to define the content of SE/PM. Much of the historical data used in this study comes from a time before the major mergers and acquisitions occurred. Some companies are attempting to integrate the accounting systems of the legacy companies and some are keeping legacy accounting systems intact. We highlight some of these differences in the accounting of SE/PM costs, but we did not try to normalize accounting systems across companies or programs.

This report deals only with SE and PM costs from the prime weapons system contractor (or contractor team) during the course of a contract. SE and PM costs also occur at the subcontractor level and within the government, which require consideration when developing life-cycle cost estimates. However, due to limitations in the availability of data, these costs are not analyzed in this report.

Organization of This Report

Chapters Two, Three, and Four provide background information on systems engineering and program management. Chapter Two defines the activities involved in SE/PM; Chapter Three explores the trends in SE/PM costs for aircraft and guided weapons programs, and Chapter Four presents the current methods used in estimating SE/PM costs. Chapters Five and Six delve into the analysis we performed in developing cost-estimating relationships. Chapter Five describes the methodology used in this study and the cost estimating relationships we developed. Chapter Six explores the effect that new acquisition approaches may have on SE/PM cost estimates. Our conclusions from this study are presented in Chapter Seven.

Appendix A provides information on systems engineering reviews and how they relate to the acquisition life cycle. Appendix B excerpts the definition of SE/PM from MIL-HDBK-881. Appendix C contains the questionnaire we used for interviewing contractors. Appendix D lists the definitions of the independent variables we that we chose to represent the potential cost drivers of SE/PM costs in aircraft programs. Appendix E provides the correlation matrices of the variables in the datasets. Appendix F describes a method for time-phasing a point estimate of a development program.

A supplement to this report (RAND TR-311-AF, not available to the general public) provides proprietary data that were used in this analysis.

CHAPTER TWO

Defining Systems Engineering and Program Management

Systems engineering (SE) and program management (PM) are important components in the development and production of complex military weapons systems. The primary focus of SE is to integrate and balance the work of engineering specialists from the initial design goals to the production of the final product. PM comprises the management (i.e., planning, organizing, directing, coordinating, controlling, and approving) of the day-to-day activities of a program as it proceeds through the acquisition process.

This chapter defines systems engineering and program management, explains the systems engineering process, and describes the tools used for managing programs and the various SE/PM tasks.

Systems Engineering Definition

SE first came into being as a separate engineering discipline during the 1950s with the Intercontinental Ballistic Missile (ICBM) program (Przemieniecki, 1993). It was first defined in a military standard (MIL-STD-499) in 1969, which was updated in draft format by MIL-STD-499(B) in 1992.[1]

[1] The draft MIL-STD-499(B) was never formally released due to the reduction of military standards (MILSTDs) and military specifications (MILSPECs) under acquisition reform. In its place, there are three prevailing commercial standards, each developed by a different industry organization: ISO/IEC 15288, *Systems Engineering-System Life Cycle Processes*; ANSI/EIA 632, *Processes for Engineering a System*; IEEE 1220, *Application and Management of the Systems Engineering Process.*

MIL-STD-499(B) defines systems engineering as follows:

> An interdisciplinary approach to evolve and verify an integrated and life cycle balanced set of system product and process solutions that satisfy customer needs. Systems engineering: (a) encompasses the scientific and engineering efforts related to the development, manufacturing, verification, deployment, operations, support, and disposal of system products and processes; (b) develops needed user training equipments, procedures, and data; (c) establishes and maintains configuration management of the system; (d) develops work breakdown structures and statements of work; and (e) provides information for management decision making.

As suggested by the above definition, systems engineering is an iterative process that involves many people from various backgrounds, not only design engineers, but also logisticians, configuration and data managers, testers, manufacturing personnel, quality control personnel, cost analysts, end users, and program managers.[2]

Many of the common SE functions found throughout programs are sometimes referred to as "ility" functions—e.g., reliability, maintainability, producibility, and survivability/vulnerability. Reliability engineers are concerned with ensuring that the system will perform as intended, without a mission critical failure, for a specified period of time. Maintainability engineers are concerned with the resource required (time, skills, and material) to repair an item if it experiences a failure. Producibility engineers evaluate the design for how well it can be manufactured using existing or new production processes. Survivability/vulnerability engineers look at "the capability of the system to avoid and/or withstand a man-made hostile environment" (Ball, 1985).

Besides these "ility" functions, SE is also composed of human-factors engineering, safety engineering, value engineering, quality as-

---

[2] The DoD definition of systems engineering has evolved somewhat to become, as stated in the *Defense Acquisition Guidebook* (December 2004), the "integrating mechanism for balanced solutions addressing capability needs, design considerations, and constraints, as well as limitations imposed by technology, budget, and schedule."

surance, corrosion prevention, life-cycle cost/design-to-cost (DTC), standardization, and other specific functional areas. Human-factors engineering relates to the performance and interaction of humans with systems. Safety engineering performs the technical analysis to evaluate the design for hazards or potential accidents. Value engineering's purpose is to achieve the required function at the lowest overall cost, sometimes resulting in what are called Value Engineering Change Proposals (VECPs). Quality assurance in SE evaluates the procedures throughout the entire process (design, development, fabrication, processing, assembly, inspection, testing, maintenance, delivery, and site installation) to ensure adequate quality. Corrosion prevention analyzes the entire design (prime contractor and subcontractor components) to determine, prevent, and control corrosion of the system in operation. Life-cycle cost analysts develop cost estimates for the program from inception through production, fielding, operations, and disposal. CAIV and DTC analysts consider the cost implications of design alternatives, oftentimes with a contractual incentive for meeting a desired unit cost goal. Standardization attempts to minimize the number of unique parts, materials, and processes within the weapon systems and within the existing industrial capability.

Systems engineering also has an Integrated Logistics Support (ILS) component that is concerned with designing the supportability features of a fielded system. One of the typically larger activities in this area is the Logistics Support Analysis (LSA). The LSA is an iterative analysis that systematically examines all the elements of a system to determine the support required to ensure the system operates effectively and keeps operating effectively. The intent is to take the ILS functions into account as early as possible to influence the design. Included in this analysis is preparation of the maintenance plan and performance of a Level of Repair (LOR) analysis to determine the optimum discard/repair levels for each hardware item. ILS systems engineering also deals with engineering change proposals; rework analysis; facilities analysis; ground support equipment management; spares and repair parts specification; training systems analysis; and packaging, handling, storage, and transportation analysis.

The Systems Engineering Process

The basic SE process translates needs or requirements into successful products and/or processes. It is largely an iterative process that provides overarching technical management of a system from the stated need or capability to an effective and useful fielded system. During the process, design solutions are balanced against the stated needs along with constraints imposed by technology, budgets, and schedules. The process is illustrated in Figure 2.1 and is described next.

Process Input

The process starts with the input of the customer's needs and requirements. The input also must include information about the desired mission or capability of the system, the technology to be used,

Figure 2.1
The Systems Engineering Process

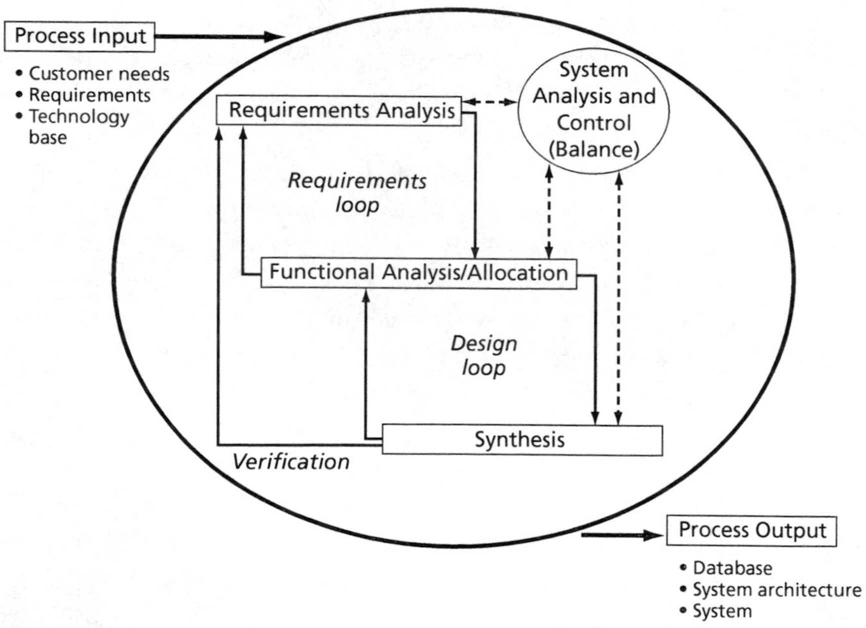

SOURCE: Defense Systems Management College, 1995.

and any constraints to the system. The input may include output from a previous phase of the process, program decision requirements, or specifications and standards to be used.

Step 1—Requirements Analysis. The first iterative step in the SE process is the requirements analysis. In this step, the systems engineer attempts to further derive, refine, and define the customer's requirements. This includes determining quantifiable characteristics that the system must possess to be successful. The relationship among requirements, the priority of the requirements, and the flexibility of the requirements are determined. The missions, threats, environments, constraints, and measures of effectiveness are continually reviewed. Some constraints that are considered are technological availability, physical and human resources, cost/budgetary impacts, and the risk of not meeting the stated schedule. The output of this process answers the question "what?" for the functionality and "how well?" for the performance requirements.

The system engineer should answer the following questions during the requirements analysis process (Defense Acquisition University [DAU], 2003):

- What are the reasons behind the system development?
- What are the customer's expectations? How will the performance of the system be measured?
- Who are the users and how do they intend to use the product?
- What do the users expect of the product?
- What is the users' level of expertise?
- What environmental characteristics (e.g., operations off of a carrier, corrosion from salt, strong electromagnetic fields) does the system have to take into account?
- What are the existing and planned interfaces?
- What functions will the system perform, expressed in the customer's language?
- What are the constraints—hardware, software, economic, and procedural—with which the system must comply?
- What will be the final form of the product—model, prototype, or mass production?

Step 2—Functional Analysis/Allocation. The second step of the SE process is the functional analysis/allocation. The goal of this part of the process is to flow down the overall system requirements to the lower-level subsystems. System-level functions include the mission, test, production, deployment, and support functions. Higher-level functions are decomposed into several sub-functions so that discrete tasks or activities can be assigned. Each function and sub-function is allocated a set of performance and design requirements, such that these low-level requirements can be traced back to the top-level requirements they are designed to fulfill. Special attention is given to making sure all interfaces, both internal and external, are addressed. Through the requirements feedback loop, the sub-functions are rolled up (re-examined) to make sure that they meet the overall requirements at the primary level.

Step 3—Synthesis. The synthesis step is the part of the process during which the allocated requirements are satisfied though design solutions at the lower levels. The collection of design solutions defines a physical architecture that satisfies the functional architecture derived in the previous step. Design requirements are set through a series of iterative design trade-off analyses. These design requirements are documented and may take the form of product baselines, product and process specifications, interface-control documents, drawing packages, facility requirements, procedural and instructional material, and personnel task loading (budgeting based on the number of people needed for each task). Through the design feedback loop, these lower-level solutions are matched back to the functional allocation that generated them to ensure that all functions are covered and all elements are justified by a valid requirement. Also at this step, the verification feedback loop ensures that the design solution meets the original requirements. Four types of verification are used to check that the design matches the requirements: examination, demonstrations, simulations and analysis, and testing.

Process Output

The final output of the entire process should be a design that fulfills the needs identified in the initial requirements established by the customer. This design is typically documented in a database, system architecture, system configuration item, or specification.

System Analysis and Control

Throughout the above steps in the SE process, system analysis and control are used to balance the requirements analysis, functional analysis/allocation, and synthesis with cost, schedule, and performance risks to ensure that the resulting system is an affordable, operationally effective, and suitable solution for the customer. Trade-off studies, effectiveness analysis, and risk management are performed toward this goal. Configuration control, interface management, and data management are important in ensuring that the entire process hangs together. Technical reviews are performed to communicate design solutions to the customer and internal stakeholders. The questions that should be asked during this part of the process and the management tools that are used to answer those questions are shown in Figure 2.2.

Figure 2.2
Management Tools for Answering System Analysis and Control Questions

Question		Management Tool
How do I make decisions?	→	Trade studies
Will it do the job? Is it worth the cost?	→	Effective analysis
Are we doing the right thing?	→	Risk management
How do I know it works?	→	Technical performance measures
Will it meet the performance criteria?	→	Modeling and simulation
Will it all work together?	→	Technical performance measures
Do we know what we have?	→	Configuration management
Are we ready to go on?	→	Technical reviews
How do I run this program?	→	Integrated planning

RAND *MG413-2.2*

SOURCE: DAU, 2000, p. 70.

Program Management Definition

Program management has been defined as "the management of a series of related projects designed to accomplish broad goals, to which the individual projects contribute, and typically executed over an extended period of time" (Wideman, 2001). Program management is very different from corporate administrative management that involves an ongoing oversight role. Program management usually has the more specific task of completing a project or set of projects for which there is a common goal and a finite termination point. The program manager has the responsibility of planning the project, controlling the project's activities, organizing the resources, and leading the work within the constraints of the available time and resources.

Project planning involves mapping the project's initial course and then updating the plan to meet needs and constraints as they change throughout the program. In the planning process, an overall plan, called an "acquisition strategy," is formulated by analyzing the requirements; investigating material solutions (designs); and making technical, cost, and performance trade-offs to arrive at the best solution. A formal acquisition plan details the specific technical, schedule, and financial aspects of a specific contract or group of contracts within a specific phase of a program. Functional plans detail how the acquisition strategy will be carried out with respect to the various functions within the program (i.e., systems engineering, test and evaluation, logistics, software development). Schedules that are continually updated are used to ensure that various milestones along a timeline are being met. Budgeting, another aspect of project planning, involves developing an initial cost estimate for the work to be performed, presenting and defending the estimate to parties responsible for budget approvals, and expending the funding.

Control of the project's activities is primarily concerned with monitoring and assessing actual activities and making sure they align with program goals. Monitoring involves conducting program reviews, measuring actual costs with planned costs, and testing incremental aspects of the program. It also includes managing the internal aspects of a program (e.g., the current contract) and monitoring external organizations (e.g., the services, OSD, or Congress) that may have a stake in the program's outcome. From time to time, a program assessment is needed to determine if the overall requirement is still being addressed, adequate funds are available, the risks are being managed, and the initial acquisition strategy is sound.

Organizing resources requires ensuring that appropriate staff members are in place to perform the activities required for a successful program. Recruiting, training, and motivating personnel are all part of the program manager's responsibilities. He or she must ensure that the organizational structure is optimized to perform the required tasks. Traditionally, programs have been organized functionally with hierarchical structures, each of which performs a certain task. Recently, IPTs have become popular for organizing personnel on a project. IPTs are multidiscipline teams with the authority and accountability to produce a specific product within a program.

Leading the work, given time and resource constraints, involves not only the previously mentioned tasks, but also directing that tasks be carried out and maintaining consensus within and outside the program. The program manager must give direction to his or her organization and take direction from organizations outside of his or her direct control. Maintaining a consensus requires making sure that the competing goals of internal and external organizations remain in balance and are working toward the desired goal.

Tasks Specific to Contractor Program Management

The weapons system contractor who ultimately performs the development and production work on behalf of the government carries out multiple tasks that are captured under program management costs. A review of WBS dictionaries, which define the work to be performed under each element in the WBS, from two programs each performed by a different contractor, revealed the following tasks in common: planning and control (also referred to as "business management"), configuration management, data management, and supplier management. Typically, there are also tasks that fall under program management that are related to performing integrated logistics support work but may be accounted for in a separate WBS element.

The planning and control task involves managing the maintenance of the program contractor work breakdown structure (CWBS), updating the project schedules, monitoring the program budget, and reporting information to the contractor's higher management and to the government customer. The CWBS details the activities that are to be performed on the entire contract and is updated as changes to the work content are agreed upon. Planning and control also involve responsibility for maintaining integrated master schedules that detail the specific work activities and show how they are dependent upon other activities performed under the contract. Budgeting and cost control are maintained using a management control system that collects cost information that is used with the schedule information to determine progress on the contract. The costs from the contract are reported to the government in cost performance reports, and the reports are used to monitor performance against both the budgeted cost and the planned schedule.

The configuration management task involves responsibility for maintaining a documentation trail of changes from the initial configuration to the final end product. This trail ensures that all required deliverable hardware and software engineering data are identified, documented, and released to the government. Configuration management must ensure that the requirements are flowed down to both the internal and subcontractor parties responsible for performing the work activities. Configuration management also includes follow-up with the release of information to make sure it is acceptable to the government client.

Data management involves administration and control of all the program data required by the customer.[3] The government negotiates with the contractor on the data that are to be delivered and formally

---

[3] For government programs, in addition to the end items to be designed or produced, lists of required data items are typically part of the contract. These items are typically spelled out in a Contract Data Requirements List (CDRL) that details the type and form of the data to be submitted and how often the reports are to be generated and delivered to the government.

makes this part of the contract as stated in the contractor data requirements list (CDRL). The contractor's duties involve establishing, maintaining, and implementing a data management program. Several current programs have established electronic technical data libraries that are used by personnel on both the government and contractor side to share information.

Supplier management involves ensuring that the items under subcontract and items that the government must supply (government-furnished equipment [GFE]) are delivered to the prime contractor. The prime contractor is the chief liaison between the subcontractors and the organizations that provide GFE to the program. This activity involves tracking and controlling items to ensure that their status is in alignment with the overall goals of the program.

The program management ILS functions include planning of the ILS program, demonstrating and evaluating the logistics program, performing a site/unit activation analysis, and preparing the ILS data in accordance with the CDRL. Much of this work includes planning, coordinating, organizing, controlling, and reporting ILS program objectives.

Tools Used in Systems Engineering and Program Management

The systems engineer and program manager's activities revolve around the basic tasks of planning, controlling, and improving a product design to meet the customer's needs. Several different tools are at the disposal of the systems engineer and program manager for developing a balanced product.

Planning Tools

In the Planning phase of a program, several tools are used to define the tasks to be accomplished in the program. Functional planning tools are also used for more-specific areas of systems analysis.

The Contract. In the planning phase, the contract provides the agreement between the government and industry contractor with respect to the system under consideration for acquisition. The contract is an outgrowth of other documents, including the request for proposal (RFP), the statement of work (SOW) or statement of objectives (SOO), specifications (SPEC), and the CDRL. The RFP establishes a need and publicly requests suggested solutions from all possible suppliers who could meet the need. The SOW is the formal statement of needs to the contractor, the SPEC establishes the technical system requirements, and the CDRL defines the data to be delivered from the contract.

The Work Breakdown Structure. The WBS is the formal definition of the work to be accomplished on the contract, and it ties the planning documents together. It takes the form of a product-oriented "family tree" (with lower-level elements summing up to form higher-level elements) that details all the hardware, software, services, and data that result from performing the SOW or SOO. Both the government and the contractor use the WBS as a tool to manage the work with a common terminology. MIL-HDBK-881 (formerly MIL-STD-881) suggests several different WBSs for various general military product categories as a guide for developing a WBS specific to the program of interest. These general WBSs can be tailored to meet the specific needs of the program.

Functional Planning Tools. Other planning tools include the Systems Engineering Management Plan (SEMP), the Software Development Plan (SDP), the Manufacturing Management Plan (MMP), the Test and Evaluation Master Plan (TEMP), the Integrated Logistics Support Plan (ILSP), and the LSA. The SEMP establishes the plan for performing the systems engineering process, including the plan for management of risk; the analysis, assessment, and verification of the work to be performed; and the design reviews and audits to be conducted. The SDP is a management plan that focuses on the software to be developed. The MMP is an execution plan that maps the transition from the planning stage to production, and it details activities such as analysis, modeling, and testing. The TEMP is a summary management document detailing the tests to be performed to ensure desired performance. The ILSP and LSA documents detail the operational support plan and support requirements.

Controlling Tools

In the controlling phase of the systems engineering process, the design is under way and the following tools provide the systems engineer with a way to maintain a logical development process through the design, build, and test stages. The tools generally fall into the categories of cost analysis, technical performance measures, design reviews/audits, and test and evaluation.

Cost Analysis Tools. There are three primary cost analysis tools. Earned Value Management (EVM) provides a means for measuring the contractor's costs and progress against the planned work schedule. Design to Cost (DTC) is a method for tracking production costs with the goal of minimizing average unit production costs while achieving the required performance. This method has in recent years given way to the Cost as an Independent Variable (CAIV) tool. CAIV also uses cost goals for a program, but considers the entire life-cycle cost and allows performance and the schedule to be traded off against cost to achieve affordability for the program.

Technical Performance Tools. Technical performance tools are measures to assess compliance with program requirements and the level of risk in a program. Performance of each element in the WBS is estimated, forecast, and assessed against the performance goals to determine how the overall system performance will be achieved. Examples of technical performance measures include weight, mean time between failures, and detection accuracy. Programs have begun identifying Key Performance Parameters (KPPs) that establish performance thresholds at higher levels. They are used to measure the success of a program's design in terms of the performance that is most important to the government customer.

Design Review/Audit Tools. Design review and design audit tools provide a periodic assessment of the design approach, the risk of the planned approach not meeting performance, cost, and schedule goals, and the program's progress toward production and a fielded system. The design review tools most frequently used include the Initial Technical Review (ITR), the Alternative System Review (ASR), the System Requirements Review (SRR), the System Functional Review (SFR), the Preliminary Design Review (PDR), the Critical Design Review (CDR), the Test Readiness Review (TRR), the System Verification Review (SVR), the Production Readiness Review (PRR), the Physical Configuration Audit (PCA), the Operational Test Readiness Review (OTRR), and the In Service Review (ISR). The design audits that are typically conducted are the Functional Configuration Audit (FCA) and the PCA that are used to verify test and production readiness. The Formal Qualification Review (FQR) confirms that the production item can be fielded. As shown in Figure 2.3, the design reviews and audits are used to evaluate performance and progress and to maintain configuration control. More details on the specific activities of these reviews and how they relate to a program's life cycle are in Appendix A.

Test and Evaluation Tools. Lastly, engineering and program management involves testing and evaluation of the designed product. Low-rate initial production (LRIP) provides a limited number of produced units and demonstrates that the manufacturing process is mature prior to beginning full-rate production. Independent Verification and Validation (IV&V) is the independent review of the functionality of the system software to ensure its effectiveness.

Figure 2.3
Program Review and Audit Process

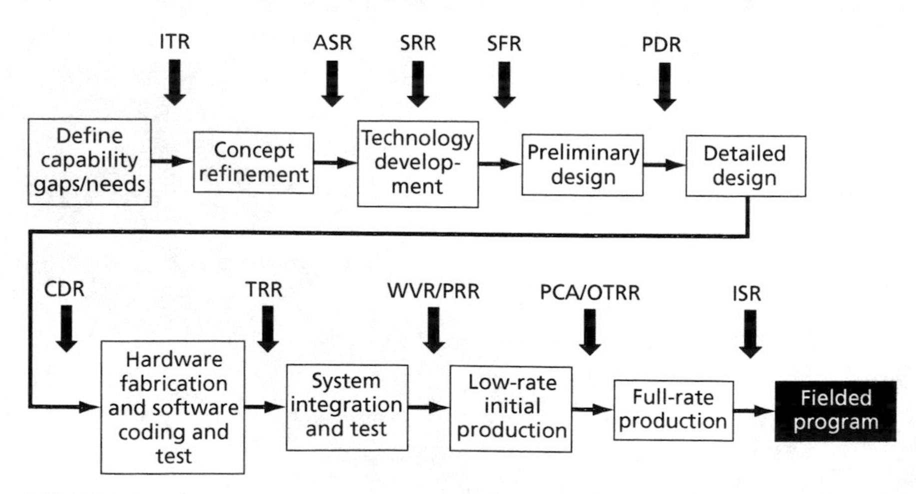

RAND *MG413-2.3*

Development Test and Evaluation (DT&E) is performed by the acquisition organization and assists the design engineers with verification of compliance with performance requirements. The government customer typically performs the Operational Test and Evaluation (OT&E) to independently verify that the final product is operationally effective and suitable. Recently, the testing and evaluation processes have become more integrated in attempt to eliminate unnecessary duplication of test activities between developmental and operational testing.[4]

Product-Improvement Tools

The product-improvement tools are concerned with efforts to further refine the product design to meet new or additional requirements. The product improvement phase generally occurs after an initial version of the product is in production or is fielded. Engineering Change Proposals (ECPs) are modifications of the end item during development or production. Post-Production Improvement (PPI) is a modification to the system in the operational and support phase after delivery of the item has been accepted. Preplanned Product Improvement (P3I) is a phased approach that seeks to satisfy incrementally different requirements relative to the priority of the government customer or the urgency of the need for the item. Evolutionary Acquisition (EA) is the acquisition approach currently favored by DoD; it recognizes the need, up front, for future improvements that will be necessary to improve capability. EA is designed to deliver incremental or spiraling capability, with the understanding that future improvements in capability will be incorporated into future design releases.

The next chapter delves into the cost aspects of SE/PM for aircraft and guided weapons programs. Cost trends for aircraft and weapons in development and in production are investigated. Some of the recent changes in the DoD's acquisition philosophy that may affect SE/PM costs are also discussed.

---

[4] See Fox et al., 2004, pp. 16–20.

CHAPTER THREE

Cost Trends in Systems Engineering and Program Management

This chapter examines the trends that have affected SE/PM costs for military aircraft and guided weapons programs. We illustrate contractor SE/PM costs over time for aircraft and guided weapons systems in development. We also show what the typical cost split is between these two categories. Historical SE/PM production cost information is presented to show the variability of the data and the overall trends in average cost throughout the production process. The chapter concludes with a discussion of the new acquisition approaches that may affect how SE/PM costs should be estimated for future programs.

SE/PM Development Cost Trends

One clear trend that is evident is that SE/PM costs are increasing in absolute terms for military aircraft development programs. Figures 3.1 through 3.8 depict the average cost for SE/PM activities in aircraft development programs grouped by decade (contract award dates) from the 1960s through the 1990s. The figures also note the number of programs for each time period. The data set includes contracts for Demonstration and Validation (D&V), Full-Scale Development (FSD) or Engineering and Manufacturing Development (EMD), and developmental modification programs. Programs from the 1980s and 1990s showed an increasingly larger amount of the development effort expended on SE/PM activities, with the average aircraft development cost over the four decades being $441 million (in fiscal year 2003 [FY03] dollars), as shown in Figure 3.1.

If the SE/PM costs are divided by the total cost of a contract, SE/PM represents about 12 percent on average of the total cost of aircraft development programs. Figure 3.2 shows that SE/PM represents an increasingly larger percentage of the cost of the development programs.

The data used for this analysis include data on a wide variety of programs ranging from tactical aircraft platforms, transport aircraft, electronic/surveillance aircraft, and bombers. An examination of the data showed that some individual programs had relatively large or small amounts of SE/PM costs and may have upwardly or downwardly impacted the average for a particular decade. Two bomber programs (the B-1 and B-2) were awarded in the 1980s and were

Figure 3.1
Trend in Aircraft SE/PM Costs for All Aircraft Development Programs, 1960s–1990s

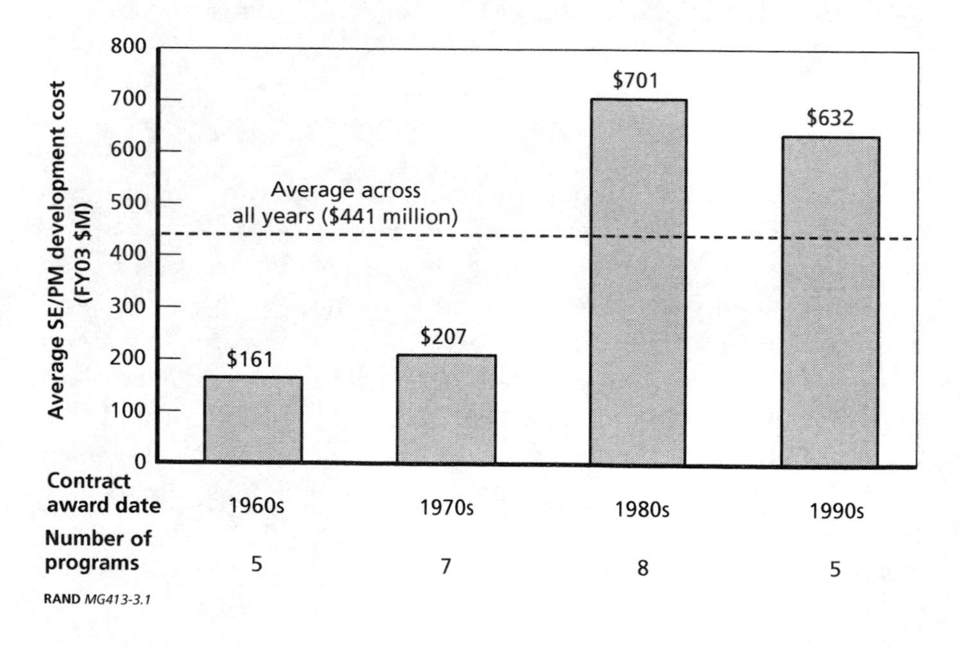

RAND *MG413-3.1*

Figure 3.2
Aircraft SE/PM Costs as a Percentage of Total Development Cost for All Development Programs, 1960s–1990s

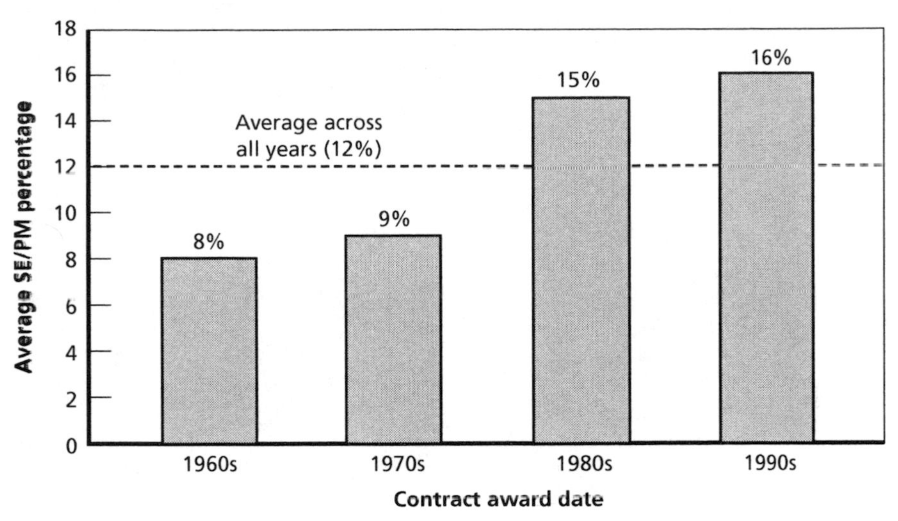

RAND *MG413-3.2*

much more expensive than the fighter/attack aircraft development programs. Also, the inclusion of D&V programs such as the YF-16 and YF-17 programs during the 1970s has the effect of reducing the average SE/PM cost, because both the total size and the proportion of SE/PM costs of the D&V programs were smaller than those of the FSD programs. For Figures 3.3 and 3.4, we eliminate the bomber and D&V programs and concentrate only on smaller-sized aircraft that experienced traditional development programs.[1] After eliminating these outlier programs, the figures still show a trend of increasing SE/PM costs, although the average cost is lower. This trend is true on an absolute basis (see Figure 3.3) and as a percentage of total program cost (see Figure 3.4). The large change in the 1970s and 1980s data, however, is due to the deletion of the bomber and D&V program data.

[1] For this subset of programs, we included recent aircraft from the following mission areas: fighter, attack, electronic warfare, antisubmarine warfare, trainer, and transport. The costs from these programs came from the larger development contracts.

Figure 3.3
Trend in Aircraft SE/PM Costs Minus Outlier Development Programs, 1960s–1990s

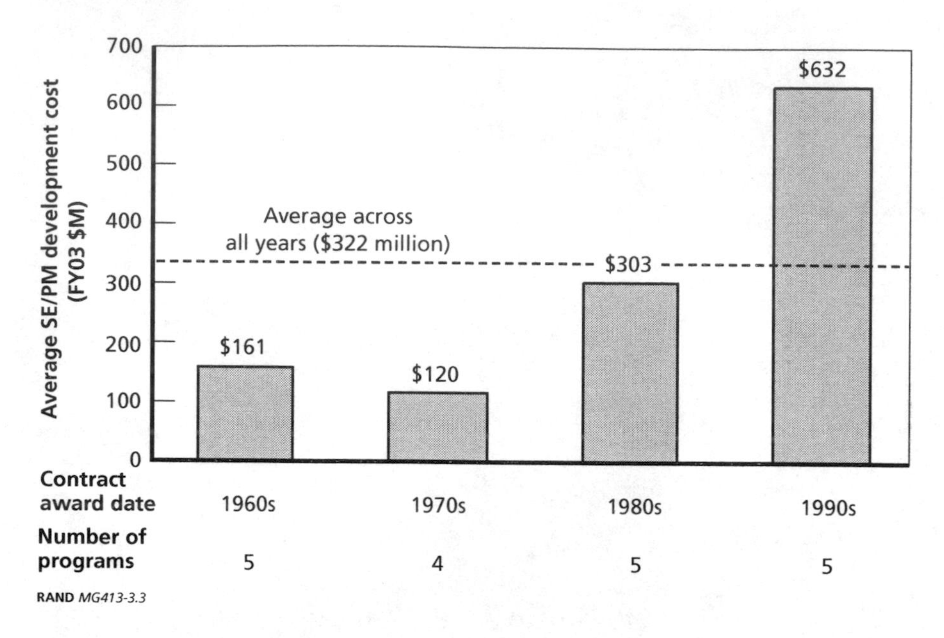

RAND *MG413-3.3*

In the case of development programs for guided weapons, there was an initial decrease in absolute cost from the 1960s to the 1970s, followed by a flat trend continuing to the 1990s (see Figure 3.5). The 1960s data point may be relatively large because it represents the average of only three weapons programs, one of which is an air-launched nuclear weapon;[2] the data points for the later decades represent a larger number of programs. The more recent programs may also exhibit lower SE/PM costs because several of them are modifications of older designs. Also, for the more recent past, we included prototype development efforts in the database and some programs that never made it into production. Figure 3.6 further shows that as a

[2] The nuclear weapons program incurred the highest SE/PM cost of all the programs in the dataset. However, eliminating this program did not change the overall trend of SE/PM costs from the 1960s to the 1990s.

Figure 3.4
Aircraft SE/PM as a Percentage of Total Aircraft Development Cost Minus Outlier Development Programs, 1960s–1990s

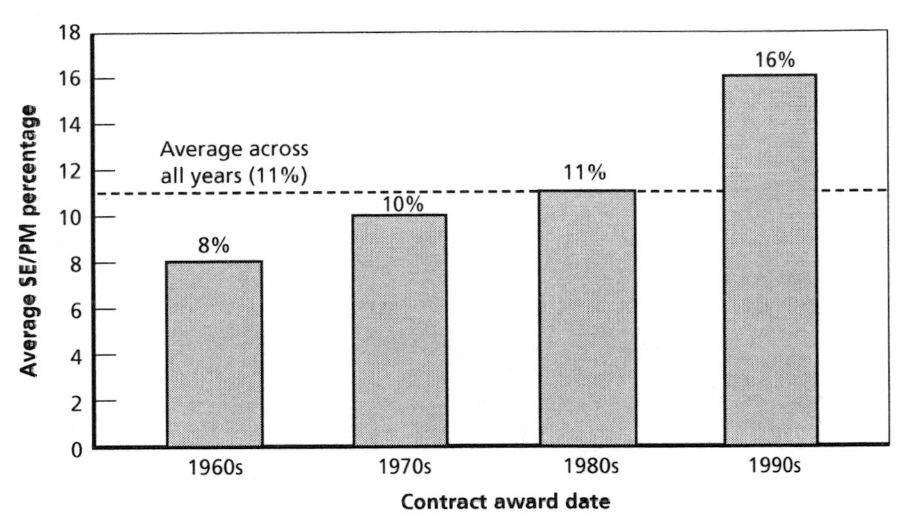

RAND *MG413-3.4*

percentage of total guided weapons program development cost, SE/PM averages 28 percent of the program cost, more than double the average for aircraft programs.

If we include only the programs that could be considered first-time development FSD/EMD guided weapons programs and eliminate prototype development programs, programs that were never fielded, and pre-planned product improvement (P3I) programs, the 1980s and 1990s saw an increase in SE/PM costs as compared with the 1970s. Also, the average SE/PM cost for this subset of guided weapons development programs shown in Figure 3.7 is about 1.5 times higher compared with the average of all the programs shown in Figure 3.5.

Again, if we normalize the data by dividing the SE/PM costs by the total contract cost for a weapons system, we see the relative change in the amount of the development effort expended on SE/PM activities. Figure 3.8 shows that there is not much difference between

Figure 3.5
Trend in Guided-Weapons SE/PM Costs for All Guided Weapons Development Programs, 1960s–1990s

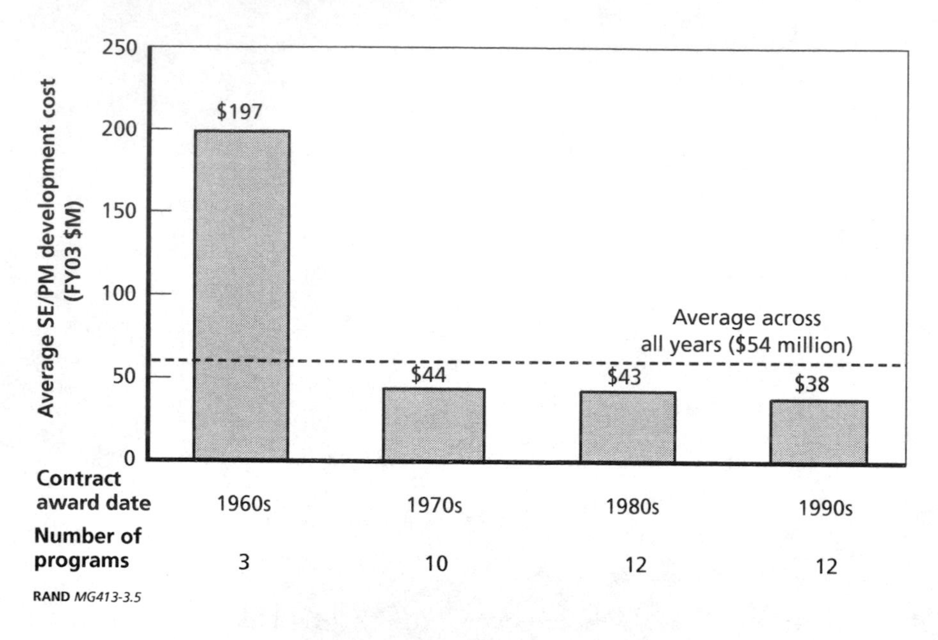

RAND *MG413-3.5*

the SE/PM costs as a percentage of development costs for this subset of programs and the SE/PM costs as a percentage of development costs for all programs (see Figure 3.6).

From the data we gathered, we were not able to derive a detailed cost breakout of the SE/PM WBS element. However, for some programs, we did have a split between SE and PM costs and some insight into SE/PM ILS costs. Figure 3.9 shows that for development programs, there is generally a 50/50 split between systems engineering and program management costs for aircraft programs and a 60/40 split between systems engineering and program management for guided weapons development programs. For aircraft programs, ILS

Figure 3.6
Guided Weapons SE/PM as a Percentage of Development Cost for All Guided Weapons Programs, 1960s–1990s

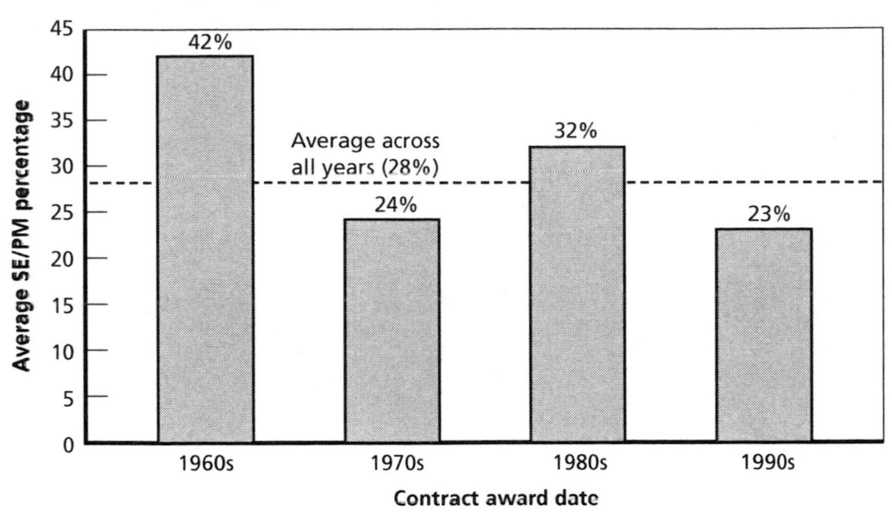

RAND *MG413-3.6*

did not make up a large part of the SE/PM costs.[3] For aircraft production programs, the data were not detailed enough to determine the split between SE and PM.

For guided weapons development programs, the split between SE and PM was more heavily weighted to SE. Also, air-to-air missile development programs had a higher portion of costs in SE than did air-to-ground programs. Air-to-air guided weapons development programs had a 69/31 split between SE and PM, while air-to-ground programs averaged a 55/45 split. As with the aircraft data, the guided weapons production cost-history data did not show enough detail to produce cost breakouts below the total SE/PM cost element.

---

[3] MIL-HDBK 881 generally suggests using separate WBS elements to collect ILS costs such as training, "peculiar" support equipment, common support equipment, operational site activation, industrial facilities, and initial spares and repair parts. The portion of ILS effort under SE/PM is related to the planning and controlling of the ILS activity and the development of the LSA plan.

Figure 3.7
Trend in Guided Weapons SE/PM Costs for FSD/EMD Development Programs, 1960s–1990s

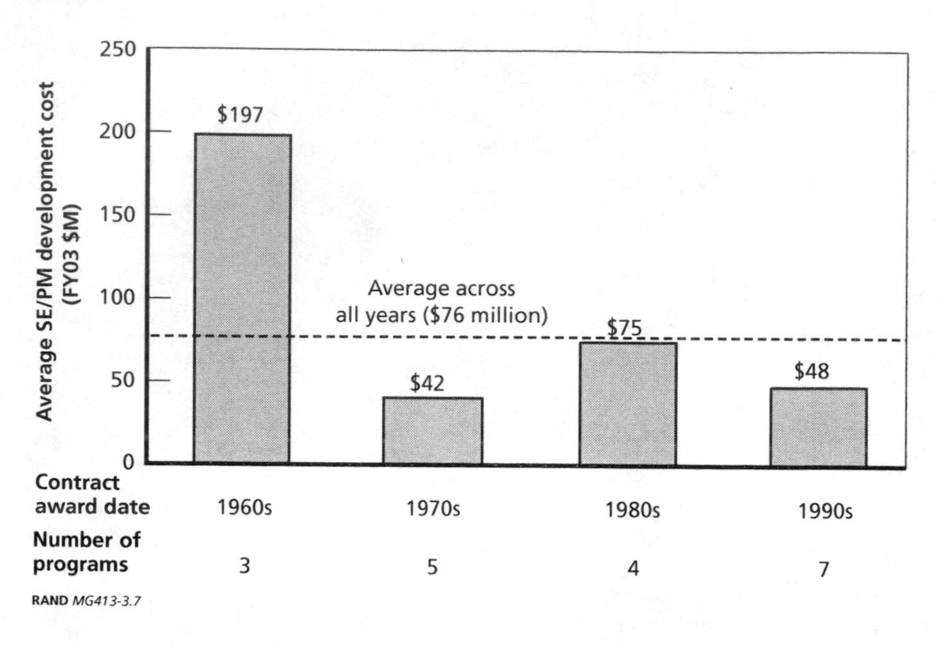

RAND *MG413-3.7*

SE/PM Production Cost Trends

After examining several datasets on production lots, we found that aircraft and guided weapons programs exhibited a large degree of variation in SE/PM as a percentage of air vehicle production costs. Figure 3.10 shows that aircraft programs display a large amount of variation in SE/PM percentages across programs as well as across production lots. The dashed line in the figure shows the average SE/PM percentage of total air vehicle cost by lot for several programs, which are illustrated by the shaded lines in the figure. The overall trend seems to be larger SE/PM cost percentages for the early lots, followed

Figure 3.8
Guided Weapons SE/PM as a Percentage of Development Cost for FSD/EMD Development Programs, 1960s–1990s

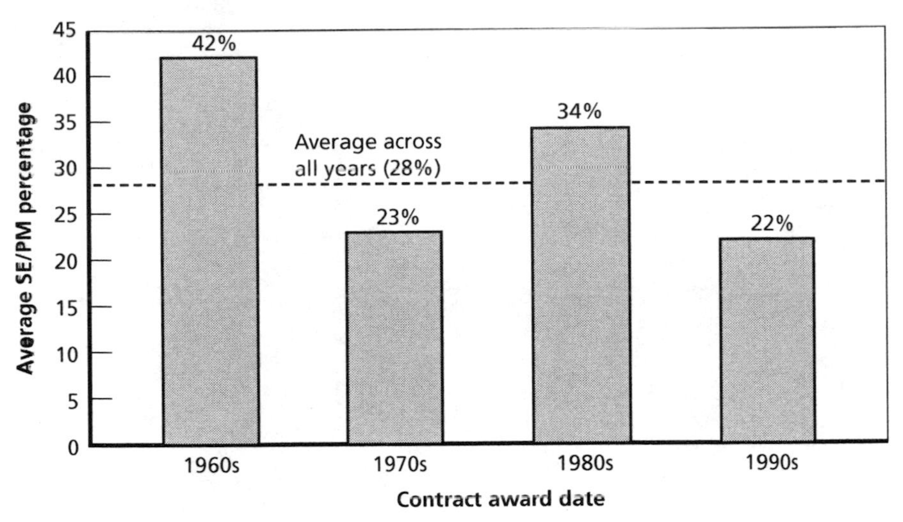

RAND MG413-3.8

by a gradual decline over the next several lots, and then a wide amount of variation for the later lots in the dataset.[4]

As can be seen in Figure 3.11, guided weapons production programs (indicated by the shaded lines) show some cost volatility in the SE/PM percentage by production lot, but the average (indicated by the dashed line) shows an anticipated decline in SE/PM as a percentage of production costs as later production lots are delivered. This may be due to the fact that weapons programs tend to run out production of a specific model until completion of a program rather than incorporate modifications into an existing production line. Usually, a new development program will include the cost of design modifications, and the production costs will restart at the top of the program's "learning curve."

[4] One of the contractors we interviewed suggested that much of the variation is due to block upgrades of the programs and the introduction of new management approaches, such as IPTs.

Figure 3.9
Average SE/PM Cost Split for Aircraft and Guided Weapons Development Programs

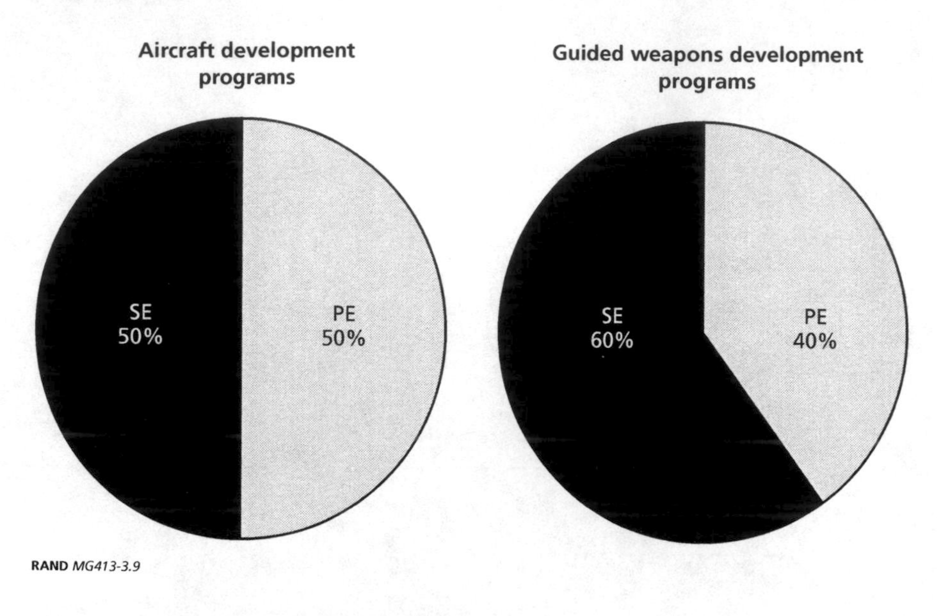

RAND *MG413-3.9*

Acquisition Initiatives That May Affect Future SE/PM Costs

During the late 1980s and 1990s, DoD pursued several strategies to try to reduce the cost of developing and procuring weapon systems, including acquisition reform. Acquisition reform is the broad array of proposals for changing the acquisition process to include more commercial-like approaches to the process and reduce the perceived burden of government regulation. RAND developed a taxonomy of reform measures in a study of the cost savings from those measures (Lorell and Graser, 2001). According to this taxonomy, the three main areas of acquisition reform are "(1) reducing regulatory and oversight compliance, (2) commercial-like program structure, and

Figure 3.10
Aircraft SE/PM as a Percentage of Air Vehicle Cost for Successive Production Programs

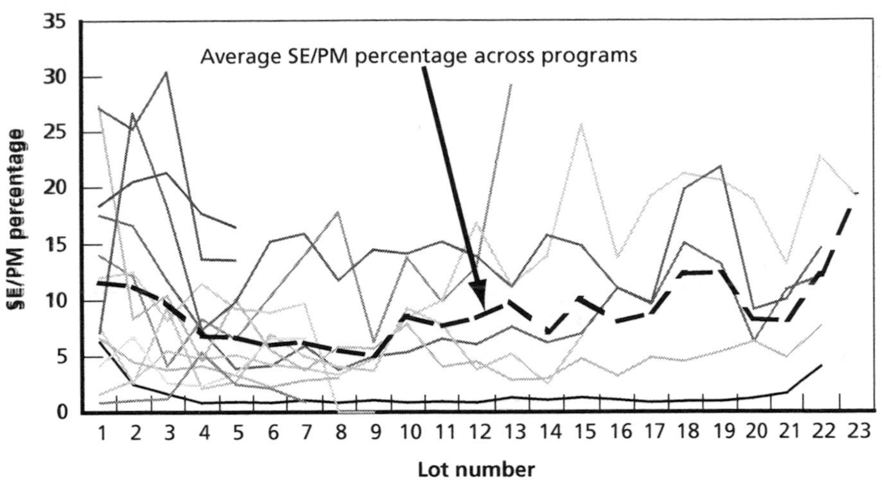

RAND MG413-3.10

(3) multiyear procurement." Only the effect of commercial-like program structures on SE/PM costs was investigated in this study.[5]

Commercial-like programs are characterized by an emphasis on concepts such as CAIV in which costs are given equal importance to performance and schedule. Levels of mission performance above the agreed-upon threshold values can be traded for cost savings if the determination is made that the extra increment of performance is not worth the extra cost. These programs can feature contractor configuration control and design flexibility to allow insertion of commercial or dual-use technology. Military specification reform and IPTs are used to implement these changes.

---

[5] The earlier RAND study concluded that relief from regulatory compliance affected overhead costs and would typically be accounted for in the labor rates used for cost estimates. The SE/PM data we used were not detailed enough to examine the effects on labor rates. Multiyear procurement savings typically would be associated only with production contracts. They are largely associated with savings at the subcontractor level and would not typically affect the prime contractor SE/PM costs we investigated in this study.

Figure 3.11
Guided Weapons SE/PM as a Percentage of Production Cost for Successive Production Lots

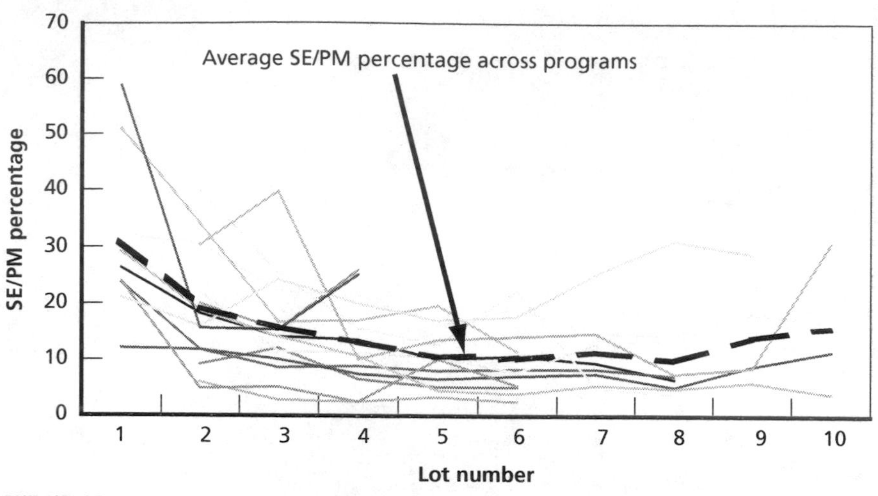

RAND *MG413-3.11*

Another area of commercial-like acquisition is the use of evolutionary acquisition. Evolutionary acquisition has recently been promoted in an attempt to reduce the time required for systems to be produced and fielded. The 5000 series[6] DoD Directives (DoDDs) recommend the use of evolutionary acquisition as a way to increase the responsiveness of the acquisition system by emphasizing the use of available technology to deliver systems in the shortest time possible. This evolutionary approach is designed to deliver capability in increments, with the upfront knowledge that future capability improvements will be necessary.[7] As stated in DoDD 5000.2, "The suc-

---

[6] Department of Defense Directive (DoDD) 5000.1 (DoD, 2003) and DoDD 5000.2 (DoD, 2003b) define management principles that should be used when managing acquisition programs within DoD.

[7] *Spiral development* and *incremental development* are two methods used to execute evolutionary acquisition. Spiral development is defined as the process by which the desired capability is known, but the end-state requirements are not known, at the start of a program. Incremental development is similar to spiral development, with the exception that the end-state requirement is known in advance.

cess of the strategy depends on consistent and continuous definition of requirements, and the maturation of technologies that lead to disciplined development and production of systems that provide increasing capability towards a material concept (DoD, 2003b)." Risk management and user feedback are used to keep the development proceeding in the right direction.

The three types of acquisition reform (reduced use of military specifications [MILSPECs] and MILSTDs, use of IPTs, and use of evolutionary acquisition) could all potentially affect SE/PM costs on a given program. Reduction in the use of MILSPECS may alleviate some of the constraints an engineer faces with a given design and thus lead to cost savings. However, this change may require some new commercial specifications to be developed and used, complicating the SE/PM work and making it more costly to accomplish. The use of IPTs could benefit the design process by including more inputs earlier in the program before the design is frozen, which could help the program avoid costs that would not have been anticipated without this forum for discussion. However, IPTs require more consensus building, which could result in more effort and time to come to agreement on a design and, thus, increase SE/PM costs in development. Evolutionary acquisition requires risk analysis to understand how the development is coming along and also requires continuous reassessment of system requirements. This acquisition approach allows for systems to be fielded with the understanding that future upgrades will be made to better meet the user's needs. However, it also requires a large amount of coordination given that each design release may resemble a "mini-acquisition" within the larger program. In summary, it is unclear if these changes would increase or decrease SE/PM costs. Each of these initiatives is discussed below.

Military Specification Reform

Since the time of World War II, the Department of Defense had been establishing a library of specifications and standards for defining components, materials, processes, testing procedures, and quality-control techniques for use in procuring defense material. The goal of developing these specifications was to develop a common terminology to facilitate communication among engineers regarding complex ideas. The government used these specifications and standards to ensure that the contractors bidding on work understood what was required and to verify that each contractor complied with the requirements of the contract.

In the early 1990s, changes in the defense acquisition environment, including a smaller defense industrial base, a rapid change in the pace of technology, and a decreasing DoD workforce, precipitated a change in dependence on military specifications and standards. As of July 1994, about 45,000 MILSPECs and MILSTDs were in existence (DoD, Office of Undersecretary of Defense [Acquisition, Technology & Logistics] 2001).[8] It was often stated that these detailed military requirements constituted a barrier to the entry of new commercial contractors and products to the military acquisition process. DoD was no longer high-tech industries' dominating buyer, and it wanted a means for accessing the most up-to-date technology in a cost-effective manner.

In June 1994, as part of the acquisition reform effort, DoD decreed an end to military standards other than performance specifications. A memorandum titled "Specifications and Standards—A New Way of Doing Business," by Secretary of Defense William Perry, cited a goal for DoD to increase its access to commercial state-of-the-art technology and allow for dual-use processes and products from the commercial sector in the military as a way to expand the potential industrial base for military equipment and services. The memo stated that this expanded industrial base would be capable of meeting defense needs at lower costs. A second goal stated in the memo was the removal of impediments to integration of commercial components into military systems. A final goal was to speed up the weapons-system development process, which typically took 12 to 15 years to field a new system. Evolutionary acquisition hopes to take advantage of what is perceived to be a faster introduction of technology, as is done in the commercial sector.

---

[8] This number reflects not only military specifications and standards but also nongovernment standards, commercial-item descriptions, and other documents.

In response to this memo, the Deputy Under Secretary of Defense (Acquisition Reform) chartered a Process Action Team to develop a strategy to reduce the reliance on MILSPECs and MILSTDs. In 1994, Secretary of Defense Perry approved the Process Action Team's recommendation "to use performance and commercial specifications and standards in lieu of military specifications and standards, unless no practical alternative exists to meet the user's needs (Perry, 1994)." Due to this policy, MIL-STD-499(A), which established the systems engineering process, was cancelled, and a final version of MIL-STD-499(B) was not released.

The Director of Systems Engineering within OSD asked that a group of organizations collaborate to develop a commercial systems-engineering standard to replace the military one. The working group was called the Electronic Industries Alliance (EIA) and was composed of members from DoD, the Aircraft Industry Association (AIA), National Security Industrial Association, Institute of Electrical and Electronics Engineers (IEEE), and the International Council on Systems Engineering (INCOSE). The resulting SE standard to replace MIL-STD-499 was called EIA-632. In parallel, the IEEE developed IEEE-1220, and the International Organization for Standardization (ISO) along with the International Electrotechnical Commission (IEC) jointly developed ISO/IEC-15288. These three commercially derived standards each address the systems-engineering process at various levels[9] and would all be needed to fully accomplish systems engineering within an organization.

The early 1990s also saw the advent of systems engineering capability-assessment models. The difference between standards and capability models is in their purpose. Standards are meant to provide an organization with processes that, if used properly, will result in an effective and efficient means for engineering a system. Capability assessment models are used to identify how well the standards are being

[9] ISO/IEC 15288 covers the life cycle of a system from concept to retirement. EIA 632 is the next lower-level standard, which addresses the processes for engineering a system (i.e., what to do). IEEE 1220 defines the process at the task or application level. See Defense Acquisition University (2004a).

implemented and point out process improvement. The primary goal of a capability model is to evaluate a program's systems-engineering capability. During the early 1990s, the Software Engineering Institute (SEI) developed various versions of the Capability Maturity Model (CMM) that was later extended to systems engineering processes.

It is difficult to determine exactly what impact the cancellation of certain military specifications and standards had on the cost of developing and procuring weapons systems. Many in industry did not anticipate the speed of the outright cancellation of many of the military standards. The government called for the creation of hundreds of nongovernment standards to be created by industrial associations, which could not keep up with the demand. In several instances, military standards were merely renamed and left intact as commercial standards.

Several pilot programs were selected to incorporate these new reform ideas into their acquisition strategies. In particular, the Joint Direct Attack Munition (JDAM) and the Joint Air-to-Surface Standoff Missile (JASSM) programs both attempted to take advantage of the new acquisition philosophy. JDAM is essentially a strap-on guidance tail kit used on "dumb" bombs, giving them much improved accuracy by using the global positioning system (GPS) and an inertial measurement unit (IMU) to provide course correction updates to the control fins during flight. JASSM is a much more complicated munition that was a follow-on program to the Tri-Service Standoff Attack Munition (TSSAM), which failed due to performance shortfalls and cost increases. These programs were some of the first to look at reducing the number of specifications by defining only a few top-level performance requirements and allowing contractors great latitude to make trade-offs to lower cost. In the case of JDAM, there were 87 MILSPECs in the predevelopment program RFP, the majority of which were eliminated by the time of the development phase (Lorell and Graser, 2001). In the case of JASSM, only three broad KPPs were stipulated to be nonnegotiable: range, missile effectiveness, and aircraft carrier compatibility.[10]

One of the objectives of this study is to determine what, if any, adjustment should be made to SE/PM cost estimates for future programs based on acquisition reform. Because both the JDAM and JASSSM programs have been through development, this report compares the SE/PM development cost of these programs to the sample of guided weapons development programs for which we collected data. We compare, in a limited way, the SE/PM costs for these streamlined programs to the SE/PM costs for programs that were developed using the conventional acquisition process.

Integrated Product Teams

Along with the introduction of military specification reform, the Secretary of Defense in May 1995 directed that Integrated Product and Process Development (IPPD) with IPTs be used throughout the acquisition process. DoD defines IPPD as follows: "A management process that integrates all activities from product concept through production/field support, using a multifunctional team, to simultaneously optimize the product and its manufacturing and sustainment processes to meet cost and performance objectives" (DoD, 1996). Rather than using a long, sequential approval process that had become the norm, IPPD attempts to drive down the decisionmaking to the lowest level possible through the use of dedicated teams of stakeholders, the IPTs.

An IPT brings together stakeholders with specific relevant experience and concerns who are committed to delivering a product to a customer. (In the case of DoD acquisition, the military user is typically considered the customer.) The expectation is that the team members will feel free to voice their concerns about a program and will be empowered to resolve issues in a cooperative fashion. If issues cannot be resolved at lower levels, the DoD model suggests that these issues be brought to a higher level for resolution. The goal of this

---

[10] JASSM added one more KPP, to address interoperability, in 2000.

framework is to bring together individuals with a wide array of expertise to work out issues jointly. Theoretically, this process should allow for a broader collection of alternatives for consideration and a quicker resolution of problems to produce a product more responsively.

As shown in Figure 3.12, the DoD IPPD framework[11] consists of IPTs at various levels that are used to execute and oversee acquisition programs. IPPD is a management philosophy that uses multidisciplinary teams composed of representatives from engineering and other technical specialties (including business and financial analysts) who along with customers achieve the best design.

Starting at the bottom of the framework, the Program IPT typically performs the execution of the specific activities on the program. The Program IPT is composed of government representatives, who support the program office, and contractor personnel, who work for the contractor that was awarded the work. The specific functional experts from both the government and the contractor are expected to work together to discuss and resolve issues in their respective areas.

The next levels of IPTs are composed of government personnel who are responsible for the oversight and review of the program.[12] Each working-level IPT (WIPT) is composed of experts in a specific field representing different organizations with different points of view. Representatives from the WIPTs participate at the next-higher level—the Integrating IPT (IIPT). An IIPT is typically a multidisciplinary group in which WIPT representatives raise higher-level issues about the program. The next level of oversight is the Overarching IPT (OIPT), whose role is to provide strategic guidance and resolve issues raised by the WIPT and IIPT. The ultimate authority for acquisition programs resides with the Milestone Decision Authority (MDA), who reviews the OIPT's input and resolves issues raised by the OIPT to enable a program to obtain approval to proceed to the

---

[11] This framework is described in DoD (1999).

[12] Industry personnel may participate in WIPT or IIPT meetings to present information, advice, and recommendations, but they are not part of government deliberations. See DAU (2004, Section 10.3.3, "Industry Participation").

next program milestone. For ACAT 1D[13] programs, the milestone decision authority is USD(AT&L) or the Assistant Secretary of Defense for Command, Control, Communications, and Intelligence (ASD[C3I]).[14]

Two programs that followed this acquisition approach in development were the F/A-22 and F/A-18E/F EMD programs. The government and the contractor worked together closely to determine requirements and incorporate changes through multiple reviews of the programs. There is considerable debate over whether the continuous involvement between the government and the contractor actually cost or saved money. In Chapter Six, we compare these programs to our sample of programs to see how their SE/PM costs differed.

Figure 3.12
Framework for DoD's IPPD Operational Structure

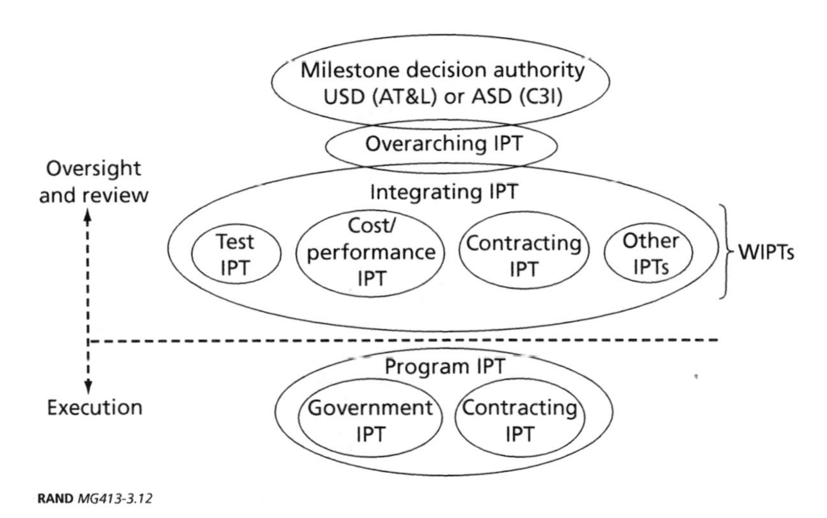

RAND *MG413-3.12*

SOURCE: DoD, 1999.

---

[13] ACAT is an acronym for Acquisition Category. The "1" indicates a major defense acquisition program (MDAP), usually defined by its cost. The "D" indicates that the MDA resides within the Office of Secretary of Defense. For an ACAT ID program, the MDA is the Component (usually military service) Acquisition Executive.

[14] The ASD(C3I) office has recently been renamed to Assistant Secretary of Defense (Networks and Information Integration) (ASD[NII]).

Evolutionary Acquisition

Evolutionary acquisition is an alternative acquisition approach, the use of which has recently been encouraged for acquiring weapons systems.[15] The impetus behind EA is that the typical serial acquisition approach used by DoD takes too long to get a program through development and into the field. EA was first applied to software-dominated Command, Control, Communications, and Intelligence (C3I) systems in which change was occurring so rapidly that it was difficult to define in detail the operational capabilities before starting EMD. Also, industry technology was being driven more by the marketplace than by DoD, and the development-cycle time within industry was much shorter than the typical development time required for a DoD program. In 2003, DoDDs 5000.1 and 5000.2 formally endorsed EA as the "preferred DoD strategy for rapid acquisition of mature technology for the user," expanding its use beyond C3I programs to all DoD acquisition programs.

The EA construct attempts to "deliver capability in increments, recognizing, up front, the need for future capability improvements" (DoD, 2003b). The construct calls for consistent and continuous updating of requirements and evaluation of technologies available to meet those requirements. Both users and independent testing organizations need to be involved to provide the necessary feedback on how well industry has identified and applied the new technologies. All of these efforts need to be balanced against time, cost, and overall priorities. Each EA increment can be treated as an individual acquisition program within a larger program. Figure 3.13 shows how the EA process can be modeled; it depicts the incremental progression of the program through the release of block versions of a system and the heavy involvement of the user in the development process.

---

[15] Preplanned product improvement and block upgrade programs are precursors to evolutionary acquisition.

Figure 3.13
Evolutionary Acquisition Model

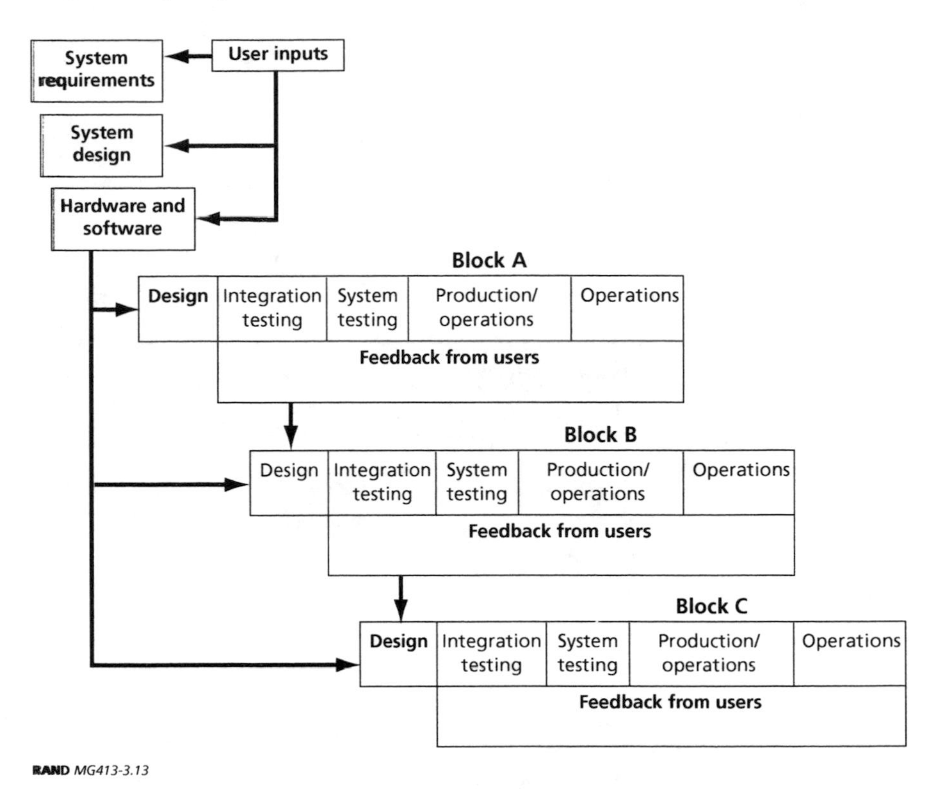

SOURCE: Defense Systems Management College, 1995.

Because the EA concept has only recently been formally applied to weapons acquisitions, the information for assessing its effect on SE/PM is limited. The Global Hawk program used EA as an acquisition technique, but the program was an Advanced Concept Technology Demonstration (ACTD) program, and historical costs were not available in the government databases used for this study. Some of the contractors we contacted suggested that it would be useful to consider P3I development programs as surrogates for EA programs. One such program is the Joint Standoff Weapon (JSOW). The program sought to develop a common "truck" missile airframe that was concurrently modified to accommodate new sensors and payloads. We used cost data from the JSOW program as a surrogate for EA to attempt to determine the impact on SE/PM cost estimates.

Summary

SE/PM represents a significant portion of development and production costs for both aircraft and guided weapons systems. Since the 1960s, SE/PM costs for aircraft development programs are increasing, while they are holding steady for guided weapon development programs. The split between systems engineering and program management costs is roughly 50/50, with guided weapons programs costs showing a slight tilt toward systems engineering. In production, we see a large amount of variation in SE/PM costs. Changes to the design process from incorporation of new acquisition initiatives, such as military specifications and military standards reform; the use of IPTs; and the implementation of evolutionary acquisition may affect SE/PM costs.

In the next chapter, we explore methods of estimating SE/PM costs that have been used in the past by both the government and by prime-level contractors. We also highlight the findings from the interviews we conducted with contractors to determine what costs they include in SE/PM and potential cost drivers that could be used to estimate SE/PM costs.

CHAPTER FOUR

Cost Data Findings and Current Estimating Approaches

We collected information for this report primarily through two means: (1) reviewing historical cost and schedule information from government reports and some internal contractor accounting information and (2) conducting discussions with government and prime-level defense contractor personnel who develop cost estimates for programs. (A copy of the questionnaire we used for interviewing contractors is in Appendix C.) From our discussions with contractor personnel, we found that the historical cost data are not always consistent in terms of which data are reported under each WBS element. We found that the current government cost-estimating methods tend to be high level and do not explicitly estimate SE/PM costs. Industry personnel use different methods to estimate SE/PM depending on the purpose of the estimate and how much is known about the program.

Cost and Schedule Data Sources

We collected the cost data from a variety of sources. For most aircraft development programs, we obtained copies of CCDRs or CPRs from the Defense Cost and Resource Center.[1] For three development pro-

[1] CCDR reporting is required for all ACAT 1 programs by DoD Directive 5000.4M (1992) and generally uses a product-oriented WBS to categorize costs. The goal of collecting cost data according to this WBS is to create a database of common cost information across different programs for systems within the same commodity (e.g. aircraft, guided weapons). The CCDR also separates the nonrecurring and recurring costs typically associated with a contract. The reporting requirement is flowed down to supporting contractors who perform a

grams, we used internal company reports. For most aircraft production programs, we used a database of costs taken from CCDRs that was developed by an analyst at the Air Force Cost Analysis Agency (AFCAA) in the late 1990s. This database provided the then-year costs for programs by major WBS elements. The database referenced the contract number and date of each CCDR used as a primary source. For most programs, the database referenced a more complete and timely set of CCDRs than we were able to obtain elsewhere. For a few production programs that were not in the database from AFCAA, we obtained copies of CCDRs. For guided weapons development and production programs, we primarily used the CCDR costs reported in the August 2000 Tri-Service Missiles and Munitions Database. For the remaining weapons programs, we obtained copies of CCDRs from other sources.

Because we observed that the duration of a contract could play a part in the analysis of cost, we obtained major milestone dates from Selected Acquisition Reports (SARs)[2] that detailed when the contract was awarded, the first flight, the end of the development test, and the first production delivery. We supplemented the data found in the SARs with schedule data found in external open-source publications. The schedule information was used to determine the duration of the development contract by measuring the number of months from contract award to each of the major milestones.

---

significant portion of the work. CPR reporting is used during the execution of the contract to measure the status of the work being performed and to document program expenditures. Unlike the CCDR, it does not separate the nonrecurring and recurring costs, but otherwise it does typically use a similar WBS cost breakout.

[2] SARs are status reports provided to Congress that are required by Title 10, USC 2432, for Major Defense Acquisition Programs (MDAPs). They provide information that covers program background, schedule, performance characteristics, funding summaries, and top-level contract information on the program. The schedule shows major design milestones within the program as it progresses from development to production and fielding. The funding summaries show the yearly amount of funding for the various appropriations that are used for the program. Typically, SARs are produced at the beginning of development and then annually at the end of the calendar year, but they may be required to be produced more frequently if there are significant changes in a program.

Cost Data Findings

For several decades, contractors were required to report costs to the government using a variety of standard reports. The primary government reports are the CCDR and the CPR. For the government to standardize cost reporting for developing a cost database, a standard WBS is used to categorize costs. MIL-HDBK-881 (DoD, 1998) provides the sample WBS by commodity (e.g., aircraft, ship, missile) that can be tailored for the specific content of the program.

The source of the cost data that feed into the government reports is the contractor's accounting system. Because the cost accounting structure of most contractor's accounting systems will not match the approved program WBS, costs are allocated or grouped into the government-approved program WBS for reporting purposes (referred to as a "cross-walk" between the accounting system and the cost report). Because of the long-standing requirement for cost reporting, the system produces a useful high-level database of program costs for many systems. The fidelity and consistency of cost reporting tend to decrease at lower levels of the WBS and for smaller-sized programs. Because systems engineering and program management require multiple disciplines to be involved in the evolution of a program's design, making sure that the appropriate efforts are tracked and accounted for can be difficult. We started with the government's definition of SE/PM costs, found in MIL-HDBK-881, and tried to determine how consistent various military contractors are in how they define SE/PM in their accounting systems.

MIL-HDBK-881 Definition of SE/PM

As described above, MIL-HDBK-881 (DoD, 1998) provides a guide for developing a WBS for detailing the work to be performed on a program.[3] It states that SE includes "the overall planning, directing, and controlling of the definition, development, and production of a system or program including supportability and acquisition logistics,

[3] The complete MIL-HDBK-881 excerpt that defines the content of systems engineering and program management costs is in Appendix B.

e.g., maintenance support, facilities, personnel, training, testing, and activation of a system. [It] *excludes* systems engineering/program management effort that can be associated specifically with the equipment (hardware/software) element." MIL-HDBK-881 defines PM as "the business and administrative planning, organizing, directing, coordinating, controlling, and approval actions designated to accomplish overall program objectives which are not associated with specific hardware elements and are not included in systems engineering."

The difficulty with both of these definitions is the stated exclusion of actual design/production engineering and management directly related to the WBS element with which they are associated. In practice, this definition is difficult to apply consistently within a product commodity, within a company, or for an entire program. Systems engineering is so intimately tied to hardware and software design that it is difficult to determine how to separately account for each. Also, with the flow down of requirements to subcontractors, some systems engineering is also done at the subcontractor level. Finally, following the mergers and acquisitions that occurred in the defense industry during the 1990s, each of the major DoD prime contractors is composed of several previously stand-alone companies. This change, along with continuing pressure to reduce overhead, has resulted in accounting systems that are continually changing at each of the major contractor organizations.

Definitions Across Multiple Contractors

To try to understand the differences in the definitions of Systems Engineering and Program Management, we gathered information from contractor proposals, WBS dictionaries, and discussions with various contractors. To get a cross section of the industry, we obtained a representative detailed subtask breakout for the major suppliers of aircraft and weapons. Table 4.1 compares the subtasks that various contractors (companies) report under SE and PM.

Table 4.1
Comparison of Contractors' Systems Engineering and Program Management Subtasks

	Contractor A	Contractor B	Contractor C	Contractor D
Systems Engineering Subtasks				
System safety	X	X	X	X
Reliability	X	X	X	X
Human factors	X	X	X	
Maintainability	X	X	X	
Producibility	X	X		X
Survivability/Vulnerability program	X	X	X	
Quality assurance	X	X		X
Requirements allocation/Validation	X	X		X
Risk management		X	X	
Standards	X	X		
Electromagnetic compatibility/ Lightning strike	X	X		
DTC/LCC program	X	X		
Test equipment		X		X
Systems engineering management		X		X
Configuration integration	X		X	
Operations analysis			X	
Corrosion prevention program		X		
Mass properties	X			
Technical baseline documents			X	
Technical integration			X	
Pilot-vehicle interface			X	
Parts standardization			X	
Weapons integration			X	
Integrity			X	
Logistics support analysis	X			
Depot and intermediate rework analysis	X			
Facilities program	X			
Support equipment program	X			
Spare and repair parts program	X			
Training systems program	X			

Table 4.1—Continued

	Contractor A	Contractor B	Contractor C	Contractor D
Program Management Subtasks				
Planning and control	X	X	X	X
Data management	X	X	X	X
Configuration management	X	X	X	X
Program management	X			X
Logistics support management	X		X	
Support equipment	X		X	
Program independent analysis	X			
Technical plans and controls			X	
Drawing control and maintenance			X	
Quality assurance management			X	
Manufacturing management			X	
Integrated logistics data			X	
Training			X	
Foreign customers			X	
Program office			X	
DTC/LCC program				X
ILS demonstration and evaluation	X			

As shown in Table 4.1, there are subtask elements in common across contractors, but also many differences among them. Systems safety, reliability, human factors, maintainability, producibility, survivability/vulnerability, quality assurance, and requirements allocation/validation tend to be included in systems engineering on a consistent basis across most contractors. Some of the differences in SE tasks may be due to accounting differences among the contractors, and they may be incorporated in other WBS elements. Several detailed areas of systems engineering (e.g., test equipment, mass properties, pilot-vehicle interface, and weapons integration) potentially could be associated with hardware design and violate the MIL-HDBK-881 definition of systems engineering.

For PM subtasks, planning and control, data management, and configuration management tend to be consistently reported across the contractor sample. However, there are subtasks that are included in SE for one contractor and in PM for another. Specifically, the

DTC/LCC effort is accounted for under both SE and PM. For guided weapons programs, we also found that some of the systems engineering effort was outside the normal SE WBS cost category. The traditional SE that concerns itself with the integration of the various components of the guided weapon was accounted for under the normal SE category. A second type of SE that includes the activities for integration of the missile on the various aircraft platforms was found in a separate WBS category. This platform-integration element includes activities such as systems requirements verification, testability, human factors, environmental protection, survivability engineering, environmental qualification, aircraft integration and interface documentation, and systems validation.

Definition of SE/PM Within a Single Company

We found from our discussions with contractors that even within a single company the definition of the content of SE/PM may change from program to program. As shown in Table 4.2, data from five programs within a single company are presented. SE/PM hours were collected across four sub-areas: SE (non-ILS), SE (ILS), PM (non-ILS), and PM (ILS). We found that work in each of these sub-areas showed differing content from program to program. Further, we found that within the SE (non-ILS) sub-area, there are work hours that could potentially violate the MIL-HDBK-881 definition that excludes design and test work from the SE/PM WBS element. The elements such as strength analysis, design support, design loads analysis, and mass properties could be classified as air vehicle hardware design, and the elements such as test support and flutter and vibration analysis could be classified as system test and evaluation.

Even though there are differences among contractors and across programs in which tasks are included in the SE and PM categories, the functions they have in common represent the cost drivers of SE/PM costs. Figure 4.1 shows the cost breakout of the lower-level SE/PM tasks for a sample program.

Table 4.2
Single Contractor's SE/PM Subtasks Across Programs

	Program A	Program B	Program C	Program D	Program E	Program F
Systems Engineering (Non-ILS)						
Technical Support	X	X	X	X	X	X
Other support	X	X			X	X
Survivability/ Vulnerability	X			X	X	X
LCC/DTC program	X				X	X
Structural design criteria		X		X		
Producibility program	X				X	
Standardization program					X	X
Strength analysis		X				
Design support		X				
Test support		X				
Design loads analysis		X				
Mass properties					X	
Systems analysis		X				
Flutter and vibration analysis		X				
Configuration control	X	X		X	X	X
Reliability/Maintainability program	X	X			X	X
Other systems engineering		X		X	X	X
Air vehicle/Subsystem analysis		X	X	X		
Safety engineering		X			X	X
Human factors engineering		X			X	X
System verification	X			X		
Sustaining engineering (non-ILS)		X		X		
Quality assurance program					X	
Environmental analysis			X			
Systems Engineering (ILS)			X			
Logistics support analysis	X	X		X	X	X
Other	X	X		X	X	X
Facilities program				X	X	
Support equipment program					X	

Table 4.2—Continued

	Program A	Program B	Program C	Program D	Program E	Program F
Systems Engineering (ILS)—continued						
Spare and repair parts					X	
Training system analysis					X	
Program Management (Non-ILS)						
Planning and control/ configuration management			X	X	X	X
Other project management	X	X	X	X	X	X
Travel	X			X		
Data management					X	X
Security			X	X		
Proposal preparation		X				X
Program management/ technical/production readiness reviews			X			X
Technical exchanges			X			
Support to project management		X				
Miscellaneous project management		X				
Program Management (ILS)		X				
ILS planning and management				X	X	X
ILS demonstration and evaluation					X	
Contractor support plans					X	
Other project management (ILS)						X

Government Approaches to Estimating SE/PM Costs

For both aircraft and weapons development programs, we found that government cost-estimating techniques generally address the total nonrecurring engineering or development cost of a program and use an allocation scheme to extract the SE/PM effort for the development program instead of attempting to estimate SE/PM costs directly. For

Figure 4.1
Detailed SE/PM Cost Breakout for a Sample Program

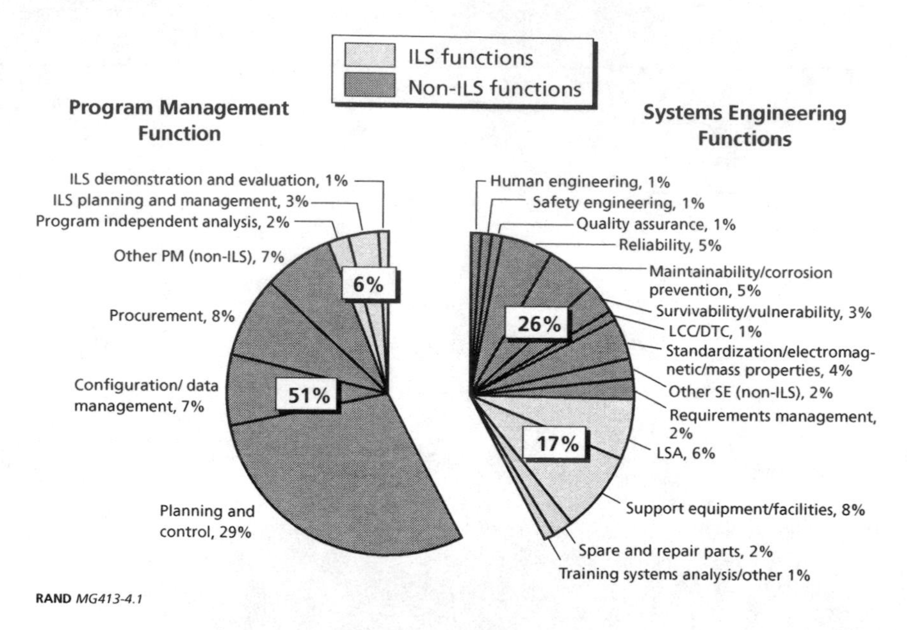

aircraft programs, the Joint Strike Fighter (JSF) program office shared their technique for estimating SE/PM costs. For missile programs, we discussed a CER used by the Naval Center for Cost Analysis (NCCA)[4] for estimating the engineering effort for development.

The JSF program uses a CER to estimate the total nonrecurring, non-test engineering (NNTE) hours based on data from six recent fighter attack aircraft development programs.[5] NNTE hours represent the total engineering development associated with the program, excluding the engineering associated with the contractor test effort. The CER parameters used as independent variables are the WE of the aircraft, the percentage of structural weight that is made of compos-

[4] NCCA was reorganized in 2002 and is now a division within the Office of Budget that reports to the Navy's Financial Management and Comptroller. It has since been renamed the Naval Cost Analysis Division.

[5] The JSF NNTE CER was jointly developed by representatives from various DoD cost agencies including the Naval Air Systems Command, the Aeronautical Systems Center, and the Air Force Cost Analysis Agency.

ites, and a "next-generation variable."[6] Based on historical ratios of SE/PM to total design engineering, an allocation of 25 percent of the total NNTE hours is allocated to SE/PM.

As with the CER used by the JSF program for estimating aircraft development engineering and SE/PM costs, NCCA regularly uses a CER for estimating total nonrecurring engineering hours for missile development. The CER uses three independent variables for determining engineering development hours: the cumulative average recurring cost of the first 1,000 units produced, the EMD phase development time, and an adjustment factor to decrement the number of hours for a product improvement program. No specific allocation is made to determine how much of this effort is associated with SE/PM.

Both the JSF and NCCA methods for estimating the nonrecurring engineering hours for an aircraft or missile development yield an estimate of the total hours including the SE/PM effort. These approaches have the advantage in that they relate the historical data from previous programs to the program of interest and make sure that all the engineering hours are accounted for in the final result. However, there is no direct estimate of the appropriate amount of SE/PM effort required.

For production programs, we found that the government uses a variety of approaches to estimate SE/PM costs. Again, early estimates on the JSF program estimated SE/PM costs as a part of sustaining engineering costs using parametric equations relating cost to such factors as weight, material complexity, and the number of units produced. It relied on cost data from five different fighter and attack programs. For a recent guided weapon program, NCCA used a factor of recurring hardware cost based on cost data from a prior program that varied from lot to lot. Both approaches used a limited amount of information (more like an analogy approach) rather than a general cross-section of program cost information.

---

[6] The next-generation variable tries to capture the additional cost of a low observable design that includes internal carriage of weapons, weight criticality, and tightly packed avionics and subsystems.

Estimating Approaches Used by Industry: General Methods

In our discussions with prime weapons systems contractors, we learned that contractors use different cost-estimating approaches depending on the phase of a program's life cycle and the purpose for the estimate. The three major phases of a life cycle are initial development, initial production, and the changes required after initial production has begun and units are fielded. The purpose of an estimate changes as a program becomes more defined. Generally, the total SE/PM cost is estimated at the aggregate level, but the lower-level tasks are tracked separately within the specific program contract.

Table 4.3 shows the various estimating approaches contractors use. The "Trade Studies" column represents programs that are very early in their conception, and estimates are required at a gross level to choose from various program alternatives. "Rough order of magnitude" estimates are performed when a desired program alternative is chosen, but a large amount of uncertainty about program specifics still exists. "Budgetary" estimates are performed when the chosen program alternative requires a greater level of detail over the performance period that the program will cover. "Firm" cost estimates are required when a contractor is providing a bid on a program for which

Table 4.3
Contractors' SE/PM Estimating Approaches

	Estimate Purpose			
Phase of Program Life Cycle	**Trade Study**	**Rough Order of Magnitude**	**Budgetary**	**Firm**
New development	Top-down	Top-down	Top-down Bottom-up	Bottom-up Top-down Combination
New production	Top-down	Top-down	Top-down	Top-down Bottom-up
Change	Top-down	Top-down Bottom-up	Top-down Bottom-up	Top-down Bottom-up

a definite scope of work has been defined. Firm estimates require the most amount of detail and the highest level of accuracy.

A "bottom-up" approach utilizes a variety of techniques based on the task, the schedule, individual judgment, and parametric approaches at a detailed level. This approach is generally used when the estimates are at a more mature stage and represent a firm quote. The "top-down" approach uses CERs at a gross or high level. The top-down approach is generally used early in a program when trade studies and rough order of magnitude estimates are required. Analogies to previous programs, in which specific comparisons with other programs are made to develop estimates for the new program, can be used for either the bottom-up (at a detailed level of cost) or the top-down (performed at a higher level of cost aggregation) approach. These analogies are used in either early program estimates or for a firm final estimate.

Estimating Approaches Used by Industry: Development Programs

To determine the methods that DoD contractors use to estimate SE/PM costs, we reviewed several recent proposals that showed various techniques that have been used on aircraft and missile programs. For development programs, three general methods are used to estimate SE/PM costs. The first uses a *constant* percentage of total or design engineering hours. The second uses a percentage that *varies* with the total design engineering hours. The third uses a level-of-effort build-up technique.

Figure 4.2 depicts variations on how SE/PM can be estimated as a constant percentage of the design effort. Each variation is differentiated by what is considered to be the "base" design effort and by what is included in the percentage applied to the base cost. The base cost is first estimated parametrically or by using an analogy to other programs deemed to be comparable to the program at hand. Each of these three constant-percentage approaches uses a percentage applied to the base cost for determining SE/PM costs. The arrows in Figure

4.2 indicate the base hours or costs that are used to multiply against a certain percentage to arrive at an SE/PM estimate.

For the NNTE allocation approach, the SE/PM hours are estimated along with engineering design and engineering technical staff hours. This method is a factored approach used in recent government estimates of aircraft development programs. NNTE hours are estimated parametrically. Using a historical percentage, the portion of the NNTE effort associated with SE/PM is determined and allocated to the SE/PM WBS element. The NNTE allocation approach acknowledges the difficulty of separating out the SE/PM effort from the overall engineering design effort by estimating all the NNTE hours together and allocating the hours back to the engineering design, engineering technical staff, and SE/PM categories.

The two other constant-percentage approaches portrayed in Figure 4.2 use the same engineering design base cost for determining the SE/PM costs, but they are used to determine SE/PM costs separately or along with engineering technical staff cost. When the SE/PM costs are estimated separately, we call it the "direct factor of design" approach. When SE/PM costs are estimated along with engineering technical staff costs, we call it an "indirect factor of design." In the direct-factor approach, the engineering design effort is estimated using parametric approaches or analogies to other programs. SE/PM costs are then directly estimated as a percentage of the design

Figure 4.2
Variations on SE/PM Cost Estimating as a Constant Percentage of the Design Effort

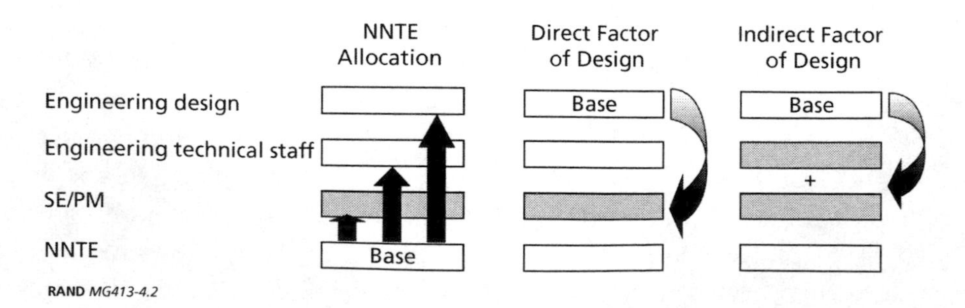

RAND *MG413-4.2*

cost. With the indirect-factor technique, SE/PM and engineering technical staff support are estimated together as a factor of engineering design hours. Engineering design hours include those associated with the wing, fuselage, empennage, landing gear, power subsystems, cooling systems, fuel systems, flight controls, electrical subsystems, cockpit systems, and weapons delivery systems. The tasks under engineering technical staff are more-general functions in which the technical engineers are experts in a specific area and support the product-specific engineers assigned to the program. Those tasks include weights/mass properties analysis, strength analysis, flight and ground loads analysis, structural dynamics, guidance and control, aerodynamics, thermodynamics, materials and processes, propulsion, loft, liaison, integrated design modeling, design quality, and observables.

The next major approach used by industry to estimate SE/PM is to use a percentage that varies according to the total design hours estimated for a program. As the design hours of a program increase, those hours as a percentage of SE/PM decrease. Figure 4.3 plots this trend (the curve) against cost data from five of the company's historical programs (SE/PM and technical support as a percentage of design

Figure 4.3
SE/PM and Technical Support as a Percentage of Engineering Design Hours Versus Actual Design Hours

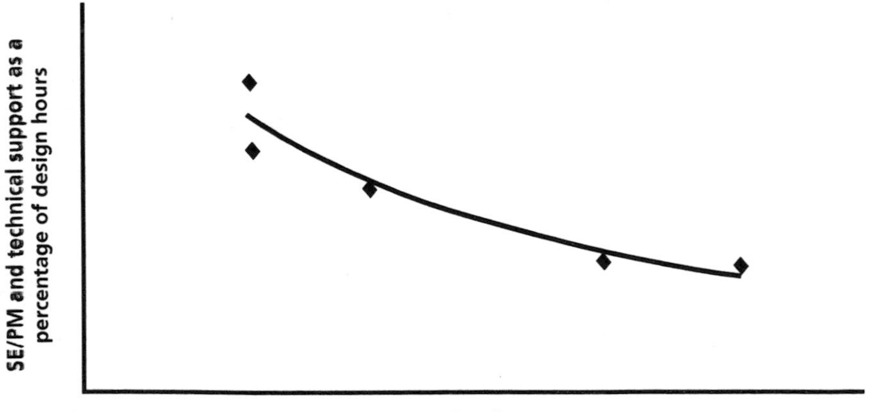

RAND *MG413-4.3*

hours versus the actual design hours incurred on a program). (Actual percentages are not shown in the figure to avoid presenting proprietary information.)

The third major approach that contractors use is a level-of-effort estimating technique, which was found in several proposals prepared for EMD contracts. This approach estimates a headcount for specific tasks based on analogies to prior programs and multiplies the headcount by an expected duration for the tasks to determine the estimated hours. This method requires detailed knowledge of both the historical program used as a basis for the estimate and the new program to be estimated. The contractors also use expert judgment and parametric approaches for estimating SE/PM.

Estimating Approaches Used by Industry: Production Programs

In production program estimates, the SE/PM engineering hours are usually part of the total production engineering hours that are estimated according to the specific type of engineering to be used. For projecting staffing for future production lots, one contractor we interviewed used a function of staffing hours versus the number of production lots. The annual number of staffing hours declined until the peak production rate was reached and then flattened for the remaining production lots. Another contractor used lower-level CERs for estimating SE/PM associated with engineering change proposals. The common base used for these estimates is the engineering design hours multiplied by a factor to account for the SE/PM costs.

Summary

Given the highly integrated nature of the work under the SE/PM category, we found differences in the definitions of what is included in the cost data for SE/PM. We found differences across contractors and within a single contractor. However, we found that the major cost drivers within SE/PM tend to be consistently reported.

Our examination of the various government and industry approaches to estimating SE/PM costs shows that different methods are used depending on the circumstances involved in each estimate. The methods used by government tend to focus on estimating SE/PM as part of the overall engineering effort and do not explicitly estimate a cost for SE/PM. The methods used by industry vary depending on the maturity of a program, the phase of the life cycle of a program, and the amount of information available about a program. Various contractors estimate SE/PM costs based on the amount of engineering design effort required. Contractors use either a constant percentage of design effort or a percentage that varies with the total design hours estimated. Level-of-effort approaches were also used to generate detailed SE/PM cost estimates.

In the next chapter, we describe various analytic approaches to estimating SE/PM costs. We focus on methods that cost analysts can use early in a program when the program's content is still being defined.

CHAPTER FIVE

Analytic Approach for Estimating SE/PM Costs

In this chapter, we discuss our approach to developing cost-estimating methodologies, the process we used to choose cost drivers, the results of our data analysis, and our recommended cost-estimating relationships. The CERs we generated estimate SE/PM contractor costs (less general and administrative [G&A] costs)[1] in constant FY 2003 dollars.

We developed a list of potential cost drivers during discussions with weapon system contractors. The list was shortened considerably when we applied two criteria to each potential cost driver: Is it readily quantifiable, and is it known and generally available to estimators early in a program? This chapter covers the estimating parameters that we chose to represent the potential cost drivers, the use and results of regression analysis to empirically test our expectations, and a recommended set of CERs for SE/PM costs in aircraft and guided weapons development and production programs. Appendix D lists these variables (i.e., cost drivers) and the definitions for each. Appendix E contains correlation matrices for the variables considered for each CER. Appendix F provides techniques for spreading a point estimate for weapons system program development costs generated by CERs to produce a year-by-year expenditure profile.

[1] We used SE/PM costs without G&A expenses because our primary data for aircraft and guided weapons are CCDRs that report costs without G&A. Also, G&A costs can vary from company to company, contract to contract, and year to year. A cost analyst using these CERs in developing estimates should account for this added expense using the specific G&A rate for the program.

Analysis of Potential Cost Drivers

We asked analysts at prime-level defense contractors what they believe are the cost drivers for SE/PM that could be used as inputs to parametric CERs. We wanted to focus on program parameters that could be associated with historical SE/PM cost data and would be available to the eventual users of the CERs. We also sought quantitative measures that were consistently reported and could be used as inputs to our CERs. The contractors' responses are summarized in Table 5.1.

As can be seen from Table 5.1, the size of a program, its duration, system complexity, technical maturity, the amount of subcontractor work, number of reviews, and level of security were all thought to be potential drivers that could affect SE/PM costs. System complexity was defined more specifically by the number of source lines of code (SLOC) developed, the number of stated requirements, the number of drawings produced, the weight of the system, and the

Table 5.1
Summary of Contractors' Responses Regarding SE/PM Cost Drivers

Suggested Cost Drivers	Contractor A	Contractor B	Contractor C	Quantification Possible?	Available to Cost Analyst?
Size of program	X		X	Yes	Yes
Duration	X	X		Yes	Yes
System complexity	X	X	X	No	No
SLOC	X	X	X	Yes	Yes
Number of requirements	X			Yes	No
Number of drawings		X	X	Yes	No
Weight	X		X	Yes	Yes
Number of subsystems (LRUs)		X		Yes	No
Technical maturity	X	X		Yes	No
Subcontract work	X	X		Yes	No
Number of reviews	X	X		Yes	No
Security		X		No	No

number of subsystems or line replaceable units (LRUs) in the platform. However, it is clear that it would be difficult to quantify some of these cost drivers, and the information might not be available to a cost estimator at the beginning of a development program (quite possibly *before* the contractor is chosen to perform the work).

Size of a program is a logical cost driver that is quantifiable and available to a cost analyst early in a program's life cycle. It can be quantified consistently by using the cost of the initial development contract. It is available to the cost analyst as a result of estimating the cost of the other various WBS cost categories and using that estimate as the cost driver for the SE/PM estimate.

Duration of a contract is another variable that has a logical relationship to SE/PM and is quantifiable and available. However, duration can be difficult to determine for historical programs. The beginning date is typically the date of the development contract award, but the end of the development program can last well into the initial production lots for a program. The SE/PM effort also stretches across the total development contract and into the early production phase.

To determine what would be a good way to measure duration, we plotted SE/PM expenditures versus the percentage of the time elapsed on various aircraft development programs and compared this expenditure profile to the occurrence of major program events. Figures 5.1 and 5.2[2] show SE/PM expenditures and total weapon system program expenditures against the program development time for two different aircraft development programs. Major program events are also shown on the horizontal axes of the figures. These figures show that SE/PM expenditures occur throughout the development program, from contract award and beyond the end of developmental test (DT End).

---

[2] The data used to create Figures 5.1 and 5.2 are derived from semiannual contractor cost-data reports from two programs. Data from six other aircraft development programs were also obtained and plotted in a similar fashion.

Figure 5.1
SE/PM Expenditure Profile over Time: Large Upfront SE/PM Effort

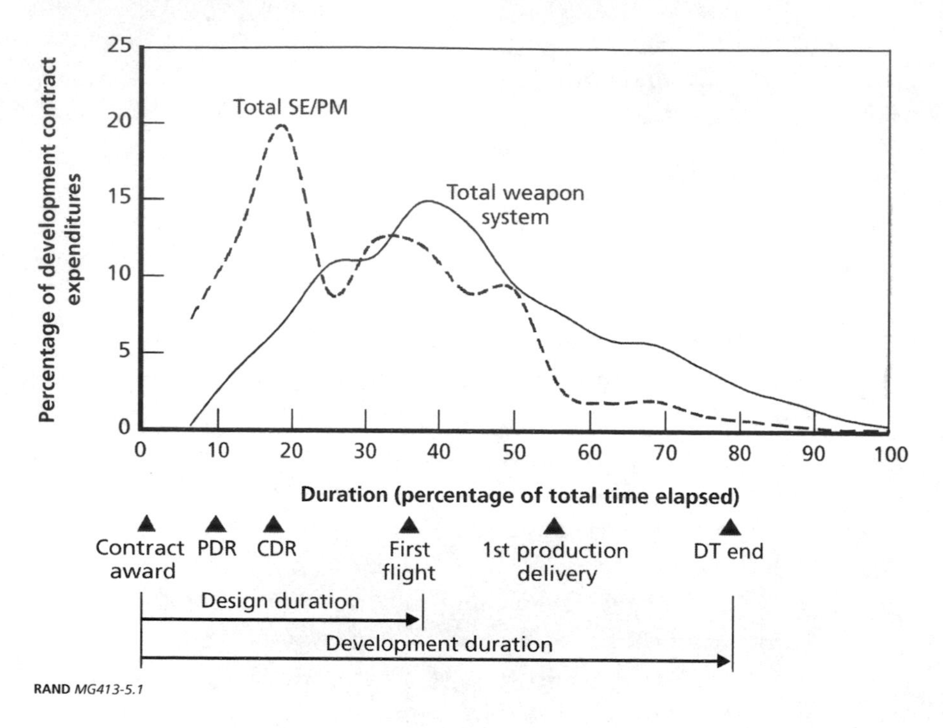

RAND *MG413-5.1*

The expenditure data we plotted fell into two general categories: high up-front expenditures for SE/PM that preceded overall program expenditures and SE/PM expenditures that tended to follow the timing of the program expenditures. Figure 5.1 further shows a proportionally large amount of SE/PM spent up front in the program as compared with the overall program expenditures for the contract. The program depicted in Figure 5.2 shows how SE/PM expenditures can closely follow the timing of expenditures over the entire program. The data we collected on six other programs evenly fell into these two categories. The figures also show that much of the effort is expended by the first flight date and the expenditures continue through the test phase. The SE/PM expenditures generally tended to trail off after DT end, making DT end a reasonable point for measuring the end of the

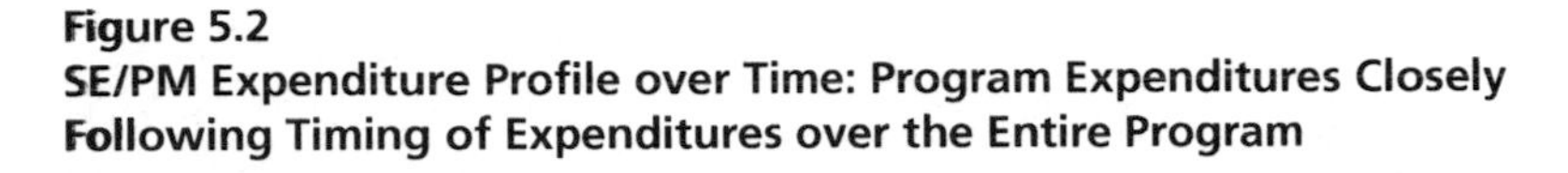

Figure 5.2
SE/PM Expenditure Profile over Time: Program Expenditures Closely Following Timing of Expenditures over the Entire Program

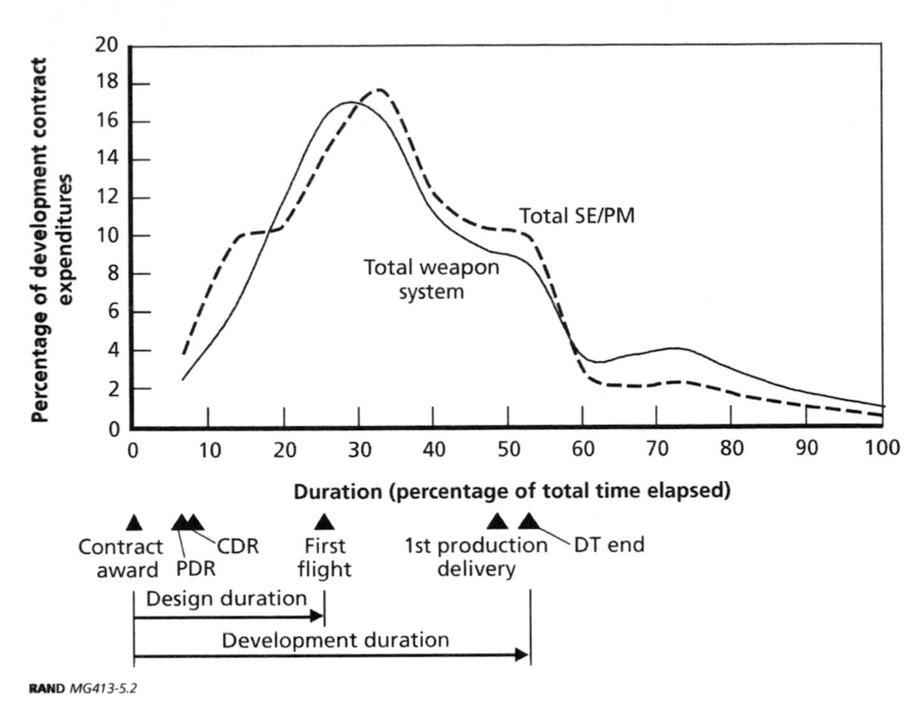

RAND *MG413-5.2*

development program. We call the time from contract award to first flight the *design duration* and the time from contract award to DT end the *development duration.* We use these two approaches to measure the duration of the development program.

Several contractors suggested that there be some means for measuring the complexity of a program. Some suggested measures of complexity were SLOC, the number of requirements, the number of drawings, the weight, and the number of complex subsystems (such as the number of LRUs). Except in the case of weight, we were unable to determine good quantitative metrics that were consistently available from historical programs and that would be available for new programs.

SLOC is a quantifiable measure available early in program development that a cost analyst could use for determining complexity of a program. However, there were some SLOC-count data available on a few of the more recent programs, but we could not find consistent data on the bulk of the historical programs in our data set. To further complicate the use of SLOC as a measure of complexity, there was no way to determine if the SLOC data we did have were consistent in the methodology used to count the SLOC.[3]

The number of program requirements is a possible measure of complexity that was suggested as a SE/PM cost driver. Although the number of requirements is a quantifiable variable, it may not be very useful for generating methods for early initial-development cost estimates. It is problematic as a measure of complexity because not all requirements take an equal amount of work to achieve, and requirements from one program to another can be very different. Recent programs have tried to use a few top-level KPPs to focus on achieving a limited set of goals rather than the many detailed specifications used for older development programs. However, these newer programs seem to show an increasing cost for SE/PM, as depicted by the cost-trend figures in Chapter Three.

The number of drawings is another measure of complexity of a program suggested by defense contractors. Although drawings are quantifiable, their use as cost drivers can be problematic. Typically, a count of the drawings is not fully determined until after a contractor has been awarded a contract to perform the development work. As with the SLOC and number-of-requirements measures, there are difficulties in consistently counting the number of drawings across programs. With the advent of computer-aided design and manufacturing and different types of drawings being generated on programs, this was not a measure we could consistently define as a means for estimating SE/PM costs.

The number of subsystems or LRUs is a factor in the complexity of technology that one contractor suggested. The subsystems and LRUs are quantifiable, but their number is generally not known until

---

[3] See Pfleeger et al. (2004), which more fully describes various methods of software sizing, including counting lines of code. The authors detail some of the inconsistencies in the methods used to count code.

much of the design has been done, and by then it is too late for early cost estimating. The number of subsystems and LRUs also was not something that could be easily determined across the set of historical programs we investigated. Use of this measure is further complicated by the degree of modularity or the addition of a weapon subsystem beyond that originally envisioned in the EMD phase.

The only complexity measure that could both be consistently defined and be available to a cost estimator at the time when initial program estimates are needed is the weight parameter. This measure has historically been one of the most widely used parameters in cost estimating and has historically been a good predictor of cost.

Technical maturity is another possible parameter suggested as a driver that could affect SE/PM costs for a program. Various measures of technical maturity have been proposed by organizations to define a quantifiable level of maturity.[4] However, technical maturity is difficult to quantify for the historical programs we studied given the current level of technology, and it is difficult to quantify for new programs without a significant amount of judgment being involved. Another problem with using technical maturity is determining how to measure it for a system such as an aircraft or guided weapons system that may have varying levels of technical maturity for various subsystems within the overall program. In the past, aircraft have benefited from ongoing research and development in areas such as materials, engines, and electronics that have occurred outside of the specific program of interest. Also, several guided weapons programs continue to focus upgrades on specific components, such as the guidance and controls, and use existing components, such as the warhead and rocket motor, from previous versions of the same guided weapon. The ongoing AIM-9 Sidewinder program is a good example of this type of continued development.

The amount of subcontract work was also mentioned as a potential cost driver for SE/PM costs. The rationale behind this suggestion is that an increasing number of subcontractors require greater

---

[4] For example, NASA uses a series of technical readiness levels (TRLs) to characterize the maturity of systems and subsystems.

coordination by the prime contractor, complicating overall planning and control and requiring additional effort to keep the entire subcontractor team updated on the evolving system design. However, this is a very difficult cost driver to measure because the amount of subcontract work to be performed is not often known at the time when initial government development estimates are required for budgeting. Changes to "make versus buy" decisions regularly occur within programs through development and production. Given these difficulties, this variable was not considered further.

The number of formal program reviews was also thought by prime contractors to be a cost driver for SE/PM costs. Again, quantification of this type of variable is not often known at the start of a program and can greatly depend on the difficulty of the individual reviews rather than on just the simple number of reviews. Also, the number of reviews of historical programs is not readily available for the programs we investigated.

The final cost driver suggested in our discussions with contractors is the level of security required for a program. The thought was that a high-security environment requires restrictions on data, personnel, and facilities that could affect SE/PM costs. However, consistent quantification of this cost driver is difficult, and this variable changes over time on a program. Also, different parts of a program may experience differing levels of security. Because of these difficulties, security level was not considered further in the quantitative analysis.

Aircraft Development SE/PM Cost-Estimating Analysis

We used the cost drivers recommended by contractors as a means for specifically defining independent variables for our analysis. We chose four general categories of cost drivers that account for the following four program characteristics: program scope, duration, physical size, and amount of integration. These general categories are further refined to specific parameters that could be quantitatively defined and compared to determine if they are good predictors of SE/PM costs.

Regression analysis was then used to determine what specific CERs were the best ones that could be developed using this analysis approach.

Aircraft Development SE/PM Cost-Estimating Parameters

As stated above, the cost drivers meeting our criteria for estimating SE/PM costs early in a program were program scope, duration, physical size, and level of integration. These four general categories were further defined quantitatively by the parameters shown in Table 5.2.

We looked at four cost parameters describing the scope of the program. The first two are similar to each other and are meant to indicate the program's scope based on the theory that the cost of a program is a good way to judge the program's size and complexity. Nonrecurring development cost (NRDEV) is the cost of the development effort less the cost of recurring items such as the cost of the developmental test aircraft. Total development cost less SE/PM (TDEVLSEPM) is the cost of the entire effort minus the cost of SE/PM. A prototype (PROTO) dummy variable identifies that a development program is an initial prototype of a demonstration program as opposed to a full-up development leading into production. The airframe already developed (AAD) dummy variable is used in

Table 5.2
Cost Drivers and Parameters Used for Aircraft Development Analysis

Cost Drivers	Parameters for Analysis
Program scope variables	Nonrecurring development cost Total development cost less SE/PM Prototype program dummy Airframe already developed dummy
Program duration variables	Design duration: months from contract award to first flight Development duration: months from contract award to end of development test
Physical size variable	Weight empty
Level of integration variable	Air vehicle cost/airframe cost

cases in which the program is still pursuing significant development, but the airframe largely has been developed prior to the program.

The next two parameters measure the duration of the development program. The time (measured in months) from contract award to first flight (CAFF) represents the duration of design. The time from contract award to end of development test (CADT) reflects the duration of total development. Figures 5.1 and 5.2 illustrate that SE/PM is expended early and throughout development programs. We chose these two measures of duration to determine whether the early part of a development program or the entire time spent in development is a better predictor of SE/PM costs.

The empty weight of the aircraft in pounds (WE) is used to see if physical characteristics play a role in estimating the SE/PM costs. WE includes the weight of the aircraft structure and its subsystems including avionics and propulsion. Weight has traditionally been used as a means for scaling the cost of development efforts when other information regarding a program is unknown. Traditionally, it has also shown a strong correlation to NRDEV.

We also tried to develop a variable to represent the amount of integration required on the program. This measure was inspired by the program integration number (PIN) used in the 1988 study by the IDA as an independent variable in estimating aircraft development costs (Harmon et al., 1988). IDA's PIN is largely a function of the amount of subcontracted effort in a development program. This variable had some appeal to us for estimating SE/PM costs because the weapon system contractors had mentioned subcontractor involvement as a driver of SE/PM costs.

However, we ended up using a different formulation of program integration for three reasons. First, the amount of subcontracted effort may not be well known early in a program. Second, our examination of the specific tasks within SE/PM and the amount of effort on each task on several historical programs showed that relatively little SE/PM effort was specifically related to subcontractors. Third, preliminary inspection and analysis of the data showed little correlation between subcontracted effort and SE/PM costs.

To represent program complexity on aircraft development programs, we calculated a factor that is the value of air vehicle cost divided by the airframe cost (AV/AF). It indicates the proportion of development effort spent on avionics and airframe. The numerator for air vehicle cost comprises the airframe, avionics, and propulsion integration[5] work required by the prime contractor in the standard WBS. Higher values indicate higher avionics content. We reasoned that programs with proportionally more effort in avionics development would require more integration effort on the avionics, and this would drive SE costs because the effort would not be associated specifically with the equipment element.

Aircraft Development SE/PM Cost-Estimating Relationships

Our next step was to use ordinary least squares and stepwise regression analysis[6] to determine which specific parameters could be used to forecast SE/PM costs for aircraft development programs. We did not include parameters in the relationships for four reasons. First, parameters were not used if they did not correlate with the dependent variable of SE/PM costs. (Correlation matrices for the parameters we considered are in Appendix E.) Second, parameters were not included if the sign of the relationships between the parameter and SE/PM

---

[5] The cost for propulsion development that included the development of the aircraft's engine was contracted separately in the databases we used.

[6] Stepwise regression is a method of selecting independent variables for inclusion in a statistical model by looking at the individual and collective contribution that adding independent variables has to explaining the dependent variable. There are various types of stepwise regression. The process we used begins with a dependent variable, but no independent variables are included. Independent variables are added one by one to the model and must result in an equation with an F-statistic that is significant at a specified confidence level (we set the level at 10 percent) to be included. The t-statistic of the added independent variable is also checked for significance. After each variable is added, the method recalculates the model with all the variables that have already been included along with the F-statistic of the equation and the t-statistic of the added variable. If the addition of any variables results in the equation having an F-statistic less than the significance level (we again set it at 10 percent), or if the t-statistic associated with the variable is less than the significance level (we used a significance level of 10 percent), it is deleted from inclusion in the equation. The stepwise process ends when all variables included in the model are significant at the 10 percent level and no variables outside the model are significant at that level.

costs did not make sense. For example, we expected that weight would be positively correlated with SE/PM costs; if the correlation matrix showed a negative correlation, the parameter would be removed from consideration. Third, we excluded parameters if they were related to other independent parameters trying to describe the same cost driver. Fourth, we excluded parameters if they were overly influenced by one or two observations that would make the parameters unrepresentative of the entire data set. Table 5.3 shows the results of the regression analysis that led to our choosing specific parameters to estimate SE/PM costs for aircraft development.

Our preferred CER for SE/PM costs in aircraft development programs is a log-linear function of duration in months from contract award to first flight and TDEVLSEPM. Although the two independent variables are related in that both are indications of the scope and complexity of a development program and are statistically correlated, the duration variable better reflects the tasks under SE/PM that act as if they are based on level of effort. The addition of the duration variable reduces the error of the model and improves the fit enough that we decided to include both variables in the model.

Table 5.3
Parameter Analysis Results for Aircraft Development Programs

Parameters	Results from Regression Analysis
Months from CAFF	Included
Months from CADT	Correlated with other variables
NRDEV	Correlated with other variables
TDEVLSEPM	Included
AV/AF cost	Not significant, wrong sign
WE	Not significant
PROTO binary or dummy variable	Not significant
AAD binary or dummy variable	Not significant

The following CER forecasts SE/PM contractor cost (less G&A costs)[7] in constant FY03 dollars. It was generated from 26 observations including aircraft prototype, modification, and full-scale development programs. The programs range from prototype programs with a few million dollars in SE/PM costs to full-scale development programs with billions of dollars in SE/PM costs.

The independent variables explain the variance in cost in the data set as a whole quite well, as indicated by the high coefficient of determination (R^2) and as depicted in the plot points in Figure 5.3 of actual costs versus predicted costs for aircraft development programs.[8] However, the extreme range of costs in the dataset results in a standard error that is relatively large for the average-sized development program as indicated by the coefficient of variation. Analysts who are estimating small development programs, such as prototype or modification programs, should be especially cautious using the following CER (in boldface), and at a minimum should cross-check its results against the costs of an analogous historical program.

SE/PM Development Cost (FY03$K) = 0.01524 * CAFF (Months) ^ 1.431 * TDEVLSEPM (FY03$K) ^ 0.7766
Adj. R^2 = 94.49 percent
F-statistic = 41.38
t-statistic of first independent variable = 2.85
t-statistic of second independent variable = 4.63
Standard error = $167.98 million

---

[7] The aircraft development programs had an average G&A percentage of 11 percent and were fairly tightly clustered around that average. The standard deviation was 2 percent with a minimum of 8 percent and a maximum of 17 percent from our dataset.

[8] We provide several statistics that can be used to assess the CERs that we generated. The R^2 adj is the coefficient of determination, adjusted for degrees of freedom. We provide the R^2 adj measured in unit space to demonstrate the model's predictive capability in showing percentage of variation explained by the regression. The R^2 adj is different when measured in log space, and it cannot be used to compare different functional forms (e.g., linear versus log-linear models). The coefficient of variation based on the standard error is a more appropriate measure for comparing across functional forms. For more information about the differences in R^2 measured in log space versus R^2 measured in unit space, see Book and Young (forthcoming).

Coefficient of variation = 39.11 percent
Number of observations = 28

For development programs, once a total estimate is made using a CER, such as the one presented above, cost analysts usually have to also look at how the estimated total cost should be spread across the duration of the development contract. Using historical cost information from aircraft development programs, we found that the SE/PM costs tend to be spread in the following manner. The first third of the cost is spent from contract award to CDR. The second third of the cost is spent between CDR and first flight of the development test aircraft. The final third is spent from first flight to the end of the development test time. We also showed that a Weibull distribution can be used to generate an expenditure profile for SE/PM costs. More

Figure 5.3
Actual Versus Predicted SE/PM Costs for Aircraft Development Programs (FY03 $)

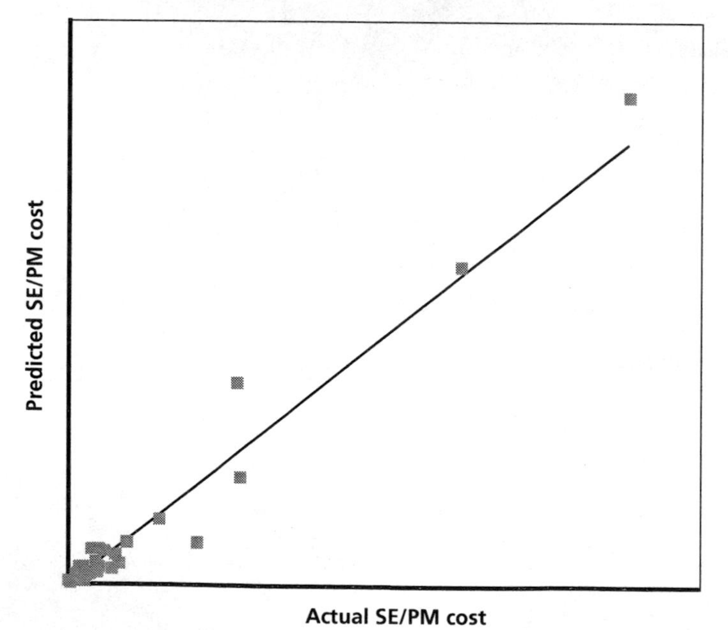

RAND *MG413-5.3*

detailed information about this profiling analysis can be found in Appendix F.

Aircraft Production SE/PM Cost-Estimating Analysis

Similar to the approach we used for developing an SE/PM CER for aircraft development programs, we investigated a similar CER for estimating SE/PM costs in production. We used the same logical approach—i.e., looking at parameters that could be used to quantify production cost drivers and using regression analysis to see how those parameters fared in predicting SE/PM costs.

Developing a production CER presented a greater challenge due to the variability of the data (see Figure 3.10). In the next subsection, we discuss out attempts to investigate the root cause of this variability to be able to model it in our resultant CER. But we were unable to accurately identify what was causing the variation. The final CERS we generated for production cost did not have a good fit to the data (as seen by low R^2 values) and should be used with extreme caution.

Aircraft Production SE/PM Cost-Estimating Parameters

We developed a different set of parameters to represent the cost drivers in production than the set of parameters representing the cost drivers in aircraft development because of the different nature of development and production programs. Unlike development programs, production programs are usually funded and executed in single-year lots. A production lot is usually awarded with deliveries starting two years after the contract award and finishing delivery by the third year after the award. Because this process is rather consistent across programs, we chose not to investigate duration of the production lot contract as a cost driver as we did for development programs.

Production costs differ from development costs in another respect that affected our list of parameters. Each aircraft program has a single full-scale development effort and cost, but each program also has several lots or years of production, each with a different cost. We tried to explain the differences in the average level of SE/PM across programs as well as the variation in costs from lot to lot on a particular program. Our conversations with weapon system contractors revealed that SE/PM costs in production are affected by changes that occur after initial development and are incorporated into each lot throughout production. For example, they mentioned that SE/PM costs are affected by the need to correct deficiencies discovered in development testing or a lot-acceptance test, by major configuration changes such as block upgrades or series changes, or by the management of simultaneous production of multiple configurations.

We considered how to formulate parameters and test the hypotheses that major configuration changes and the simultaneous production of multiple configurations would affect SE/PM costs. Almost all aircraft programs undergo some configuration changes in production. We defined major configuration changes as a model change to a different series, for example, the F/A-18A/B changing to the C/D. In our dataset, the fighter aircraft offered a subset of programs with long production histories and major changes in configuration that would allow us to test the hypothesis. The fighter programs also had foreign military sales (FMS) at various times in their history. FMS aircraft are sold in slightly different configurations than aircraft for the U.S. military and involve a different customer. These programs allowed us to test the hypothesis that management of multiple configurations would affect SE/PM costs.

We plotted the data on the fighter programs in our dataset to examine these questions and understand how we could formulate parameters to account for major configuration changes and the production of multiple configurations. Among other issues, we investigated whether an increase in SE/PM preceded the introduction of the change, was coincident with the change, or followed the introduction of the change into the production line. After plotting the data (see Figure 5.4), we determined that there was no consistent relationship between major changes in configuration and increases in SE/PM cost per aircraft. One program did show a significant increase in cost per aircraft coincident with a change in model, but the other programs showed no clear pattern. The one program with a significant increase

Figure 5.4
Production SE/PM Costs per Fighter Aircraft with Series Change

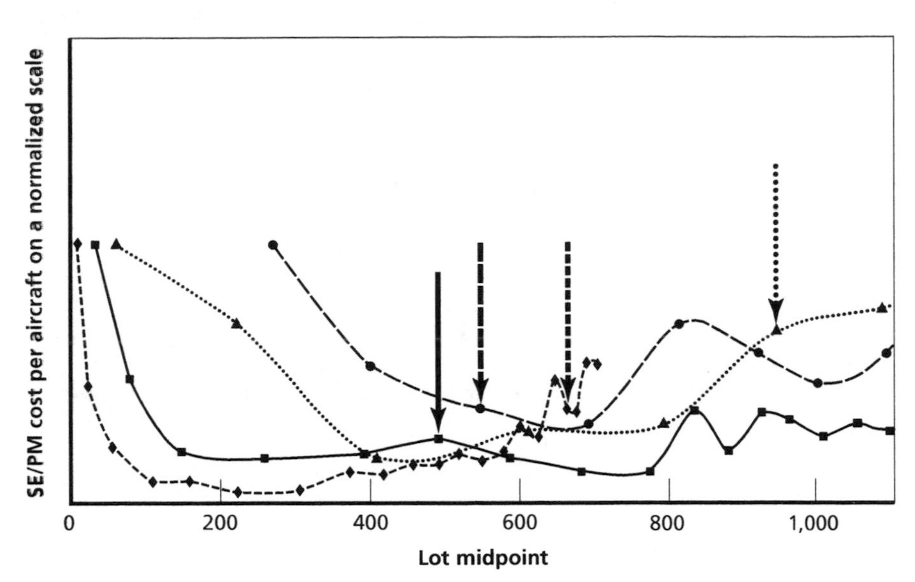

RAND MG413-5.4

NOTES: Arrows indicate last year of original series. Each curved line represents the cost data for a specific aircraft production program.

in SE/PM cost per aircraft also had a significant increase in SE/PM cost *per lot* upon the introduction of its new series that was sustained long after the change in configuration.

To better understand this unpredictable cost behavior, we looked at internal company reports that identified recurring and nonrecurring SE and PM labor hours for the one program with increased SE/PM costs. The CCDRs for the programs reported the SE/PM effort for all programs as a recurring effort, which is a typical reporting practice if most of the production effort is recurring in nature. The internal reports revealed that most of the SE/PM cost increase after the introduction of the follow-on aircraft series was for nonrecurring SE labor. Such effort may have been for the design of engineering changes associated with the model change, related studies, or nonrelated efforts. In any event the increase was due to nonrecurring tasks in production and, therefore, was impossible to forecast.

We performed the same analysis to see if changes in FMS could explain fluctuations in SE/PM production costs by lot.[9] Figure 5.5 shows the plot of yearly SE/PM costs and the years in which FMS took place. We saw no evidence that FMS increased SE/PM costs. The figure shows that for two programs SE/PM costs per aircraft were unchanged during FMS years. The third program shows no apparent relationship between FMS in general or the number of customers and SE/PM costs per aircraft.

Figure 5.5
Production SE/PM Costs per Fighter Aircraft with Foreign Military Sales

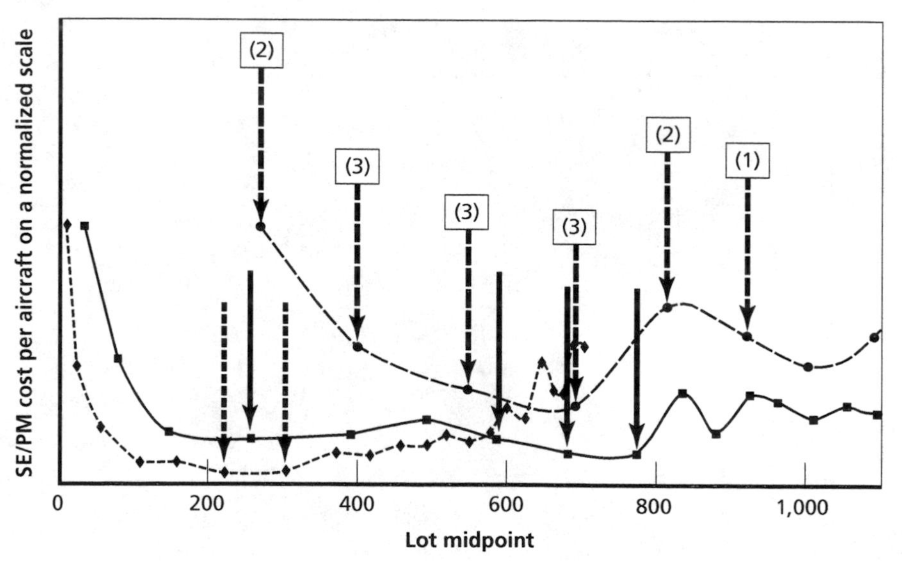

RAND *MG413-5.5*

NOTES: Arrows indicate last years of significant foreign military sales. Numbers in parentheses indicate number of FMS customers. Each curved line represents the cost data for a specific aircraft production program.

---

[9] One program shown in Figure 5.4 was removed because it had FMS throughout the production program data.

After examining the data to test our hypotheses, we found no evidence to show that configuration changes or management of multiple configurations resulted in a significant increase in SE/PM costs. Although the hypotheses are intuitively appealing, they were not born out in the data we gathered. We did not pursue the analysis any further with the complete data set.

Table 5.4 shows the parameters we chose to represent cost drivers in production.

We believed that in production, as in development, the amount of SE/PM would be related to the scope and complexity of the aircraft. The first four variables in Table 5.4 are costs to indicate the scope and complexity of the aircraft. The air vehicle average unit cost per lot (AV AUC) and air vehicle 100th unit cost (AV T100) represent the recurring cost of the airframe and contractor-furnished avionics. It does not include the cost of the engine and other cost elements that make up the unit recurring flyaway cost (i.e., SE/PM, engineering change orders, or government-furnished equipment). It generally does not include G&A and profit.

The physical variable we investigated was the empty weight of the aircraft. As in the case of development, weight is a variable that has historically been used in estimating and can serve as a proxy for complexity.

Table 5.4
Cost Drivers and Parameters Used for Aircraft Production Analysis

Cost Drivers	Parameters for Analysis
Program scope variables	Air vehicle average unit cost per lot (AV AUC) Air vehicle 100th unit cost (AV T100) NRDEV SE/PM costs in development (SEPM DEV)
Physical variable	WE
Programmatic variables	Lot number Lot midpoint rate per year Production quantity rate ratio

For representing programmatic cost drivers, we looked at four different parameters for predicting SE/PM production lot costs. We used the variables of lot number (LOT NUM) and lot midpoint (LOT MP) to reflect the expectation that declining SE effort would be required after deficiencies discovered in testing are corrected in early production lots and the aircraft configuration stabilizes. This expected decrease in effort with cumulative quantity would be offset in some years by increased effort caused by major configuration changes and production of multiple configurations, if applicable. Algebraic LOT MPs were calculated using the average slope of the entire dataset.

We included the variable of rate per year (RATE) to reflect the idea that recurring SE/PM may have some fixed costs associated with it, such that the cost per aircraft would increase for a smaller-sized lot and decrease for a larger-sized lot. The production quantity rate ratio variable is similar in intent to the rate variable, but it is calculated separately for each production lot of the program. Production quantity rate ratio is defined as the ratio of a given year's production lot quantity divided by the maximum production lot quantity achieved for that program (Q_n/Q_{max}).

The final variable we investigated was the effect of the last production lot. Looking back at Figure 3.10, it appears as though SE/PM costs tend to increase toward the end of production, and the last production lot might experience an increase in cost. To see if this variable might have an effect on cost, we introduced a last lot (LLOT) binary parameter.

Aircraft Production SE/PM Cost-Estimating Relationships

The variables that we evaluated in the regression analysis are shown in Table 5.5, along with the reason for the exclusion of any parameter from the CER.

Because of the large variation in aircraft production SE/PM costs, the CERs we generated showed poor fit statistics. Examination of the data plots illustrates the problem. Figure 3.10 presented one illustration of the unruliness of the SE/PM costs in aircraft

Table 5.5
Aircraft Production Program Parameters and Reasons for Their Exclusion from the CER

Parameters	Results from Regression Analysis/Reason for Exclusion
AV AUC	Not significant
AV T100)	Not significant
SEPM DEV	Included
NRDEV	Wrong sign
RATE	Included
LOT NUM	Wrong sign
WE	Not significant
LOT MP	Included
Production quantity rate ratio	Included
Last lot binary or dummy variable	Not significant

production, and all of the CERs we fit to the historical SE/PM costs were unable to predict SE/PM costs very well. There is large variation in SE/PM costs among programs, whether the costs are calculated per production lot, average per aircraft per lot, or per pound of empty weight per aircraft per lot. Furthermore, there is often large variation in SE/PM costs on an individual program that is not explained by the cumulative quantity or rate effect that one normally sees with production cost data. For some programs, the overall trend is a decrease in cost over time in production; for other programs, the overall trend is the opposite.

Our approach to developing CERs for production was to first generate equations similar to the functional form represented by the standard cost-improvement curve. The general functional form, also referred to as a "learning curve," is represented by the following equation: Y = A * X ^ B, where Y is the average unit cost for the Xth unit and A and B are constants.[10] The A value provides a means of scaling the curve on the Y-axis and represents the cost of the first unit. The

[10] Usually aircraft and guided weapons are bought in production lots. The X value represents the algebraic midpoint of the particular production lot for an associated Y value.

B value is a constant that depicts the change in cost as X increases from the first unit. The B value is typically expected to be a negative number less than one that gives the function an exponentially decaying shape.[11]

A variation of this basic cost-improvement curve includes an additional term to account for changes in yearly quantity. The functional form of this model is Y = A * X ^ B * Q ^ R, where Q is the yearly production rate and R is a constant. Similar to the basic equation in the previous paragraph, the R value is typically a negative value less than one that provides a means for further adjusting the unit cost down for an increase in yearly quantity, or adjusting the unit cost up for a decrease in yearly quantity.

Using the functional form including the term for yearly production rate, we produced two possible CERs that could be used to predict SE/PM unit cost in production. The first version of the CER uses the cost of SE/PM in development as a means of scaling the Y-axis starting point (or A value). We hypothesized that the amount of SE/PM expended in development should be related to the amount of SE/PM required in production. The second variation of this basic model uses the recurring cost of the air vehicle at the 100th unit (T100) as a way to scale the SE/PM production start point. The expectation was that the higher the air vehicle cost, the more SE/PM costs would be required in production.

The equation (in boldface) and the statistical results from the analysis using SE/PM development cost for scaling the first CER follow. The equation forecasts SE/PM contractor cost (less G&A cost)[12] in FY03 dollars.

Production SE/PM Cost per Aircraft (FY03$K) = 2133 * SEPM DEV(FY03$K)^ 0.1957 * LOT MP ^ –0.09178 * RATE ^ –0.768

---

[11] The B value is given as the natural log of the learning curve divided by the natural log of 2. The learning curve is usually expressed as a percentage and is equal to 2^B.

[12] The aircraft production programs had an average G&A percentage of 6 percent. The standard deviation of G&A was 2 percent.

Adj. R^2 = 51.94 percent
F-statistic = 89.86
t-statistic of first independent variable = 3.21
t-statistic of second independent variable = –1.78
t-statistic of third independent variable = –11.70
Standard error = $2.87 million
Coefficient of variation = 106.05 percent
Number of observations = 165 lots, 13 programs

The first CER that we produced has coefficients that indicate a unit learning curve of 93.8 percent and a rate curve of 58.7 percent. Figure 5.6 shows how closely the first CER predicts SE/PM production costs (represented by the diagonal line) as compared with the actual costs for the 165 observations (yearly production lots) of a

Figure 5.6
Actual Versus Predicted SE/PM Costs for Aircraft Production Programs (FY03 $ per aircraft): CER 1

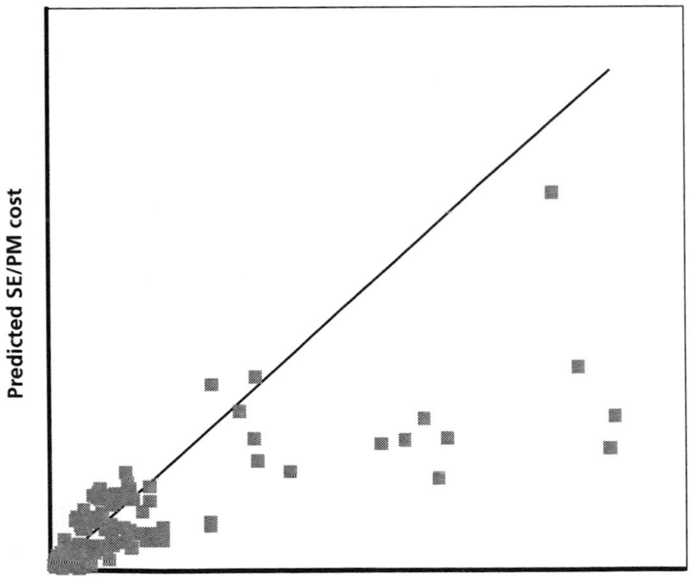

RAND *MG413-5.6*

variety of types of Navy and Air Force aircraft production programs (represented by the plotted squares). The figure shows that the CER does a poor job of predicting higher-value programs as the data points drop from the predicted line. Given this fact, the CER should be used only as a rough crosscheck, preferably in conjunction with an analogy of the same type of aircraft produced by the same contractor.

The second CER we developed with this same functional form uses the recurring air vehicle cost at T100 as a y-axis scaling variable. Using the T100 value in the equation allows the SE/PM estimating equation to be scaled on the y-axis using production recurring costs instead of SE/PM development costs. The equation (in boldface) and the statistical results for the second CER are as follows:

> **Production SE/PM Cost per Aircraft (FY03 $K) = 2682 * AV T100 (FY03$K) ^.1832 * LOT MP ^ -0.1599 * RATE ^ –0.5678**
> Adj. R^2 = 51.46 percent
> F-statistic = 82.7296
> t-statistic of first independent variable = 1.8486
> t-statistic of second independent variable = –3.1871
> t-statistic of third independent variable = –9.997
> Standard error = $2.86 million
> Coefficient of variation = 106.18 percent
> Number of observations = 165 lots, 13 programs

This second CER for production SE/PM costs per aircraft has very similar fit statistics to the first CER that used SE/PM development cost as an independent variable. The second CER has coefficients that indicate a unit learning curve of 89.5 percent and a rate curve of 67.5 percent for the same data points used to generate the previous CER. As was the case with the first CER, this CER yields a large amount of difference between the actual and predicted costs, especially for higher-cost programs, as Figure 5.7 shows.

Given the problem of the poor predictive capability exhibited by the CERs as shown in Figures 5.6 and 5.7, we tried to slightly alter

Figure 5.7
Actual Versus Predicted SE/PM Costs for Aircraft Production Programs (FY03 $ per aircraft): CER 2

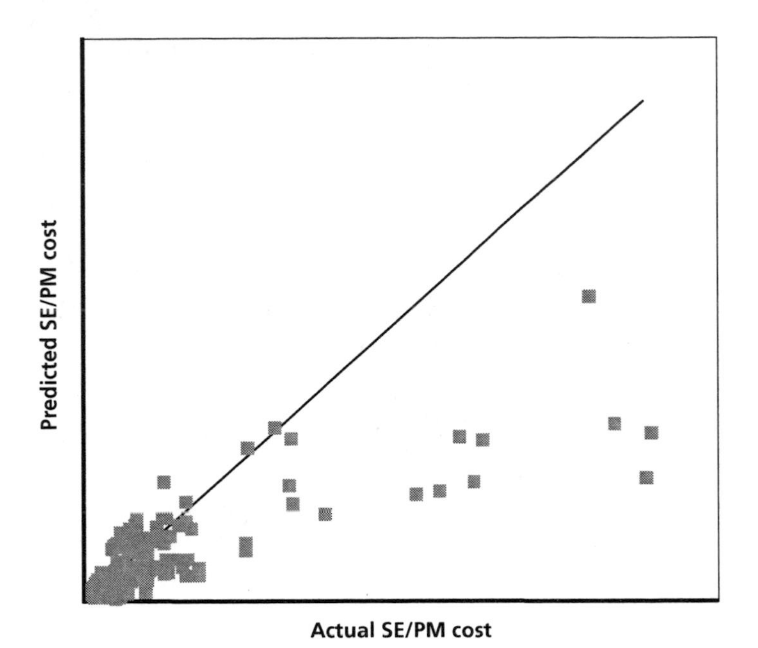

RAND *MG413-5.7*

the basic functional form and investigate the results. The functional form that we used is given by the following equation: Y = A * X ^ B * (Q_n/Q_{max}) ^ R. In this equation, the yearly lot quantity is divided by the maximum yearly production quantity.[13] This approach provides a multiplicative term that provides a penalty (a cost increase) for lots that are below the maximum quantity production lot size. We felt that this was reasonable given that many of the SE/PM functions act as fixed costs and would be minimized (from a unit cost perspective) when the most units were being produced. We further found from examining the yearly production data that this was in fact occurring for some of the programs in our dataset. Figure 5.8 depicts the SE/PM unit cost as a percentage of total air vehicle cost by produc-

---

[13] This formulation is discussed in Lee (1997, p. 60).

tion lot as compared with the corresponding Q_n/Q_{max} values for each production lot (depicted by the plot points) for a select aircraft production program.

This specific program experienced a ramp up to the maximum production rate, a ramp down, a ramp back up to the maximum production rate, and a final ramp down.

The third CER (shown in boldface) and associated statistics from the logarithmic functional form using the production quantity rate ratio are as follows:

Production SEPM Cost per Aircraft (FY03$K) = 28 * AV T100 (FY03$K) ^ 0.5216
*** LOT MP ^ –0.3435**
*** (Q_n/Q_{max}) ^ –0.4623**

Figure 5.8
SE/PM Percentage of Air Vehicle Cost Versus Rate Ratio

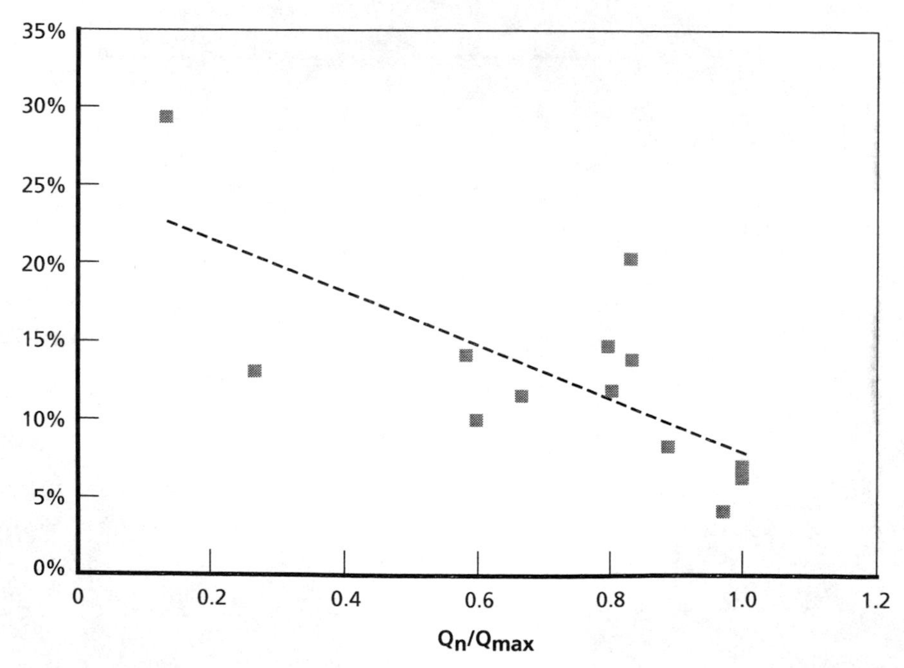

RAND *MG413-5.8*

Adj. R^2 = 44.51 percent
F-statistic = 50.52
t-statistic of intercept = 2.57
t-statistic of first independent variable = 4.72
t-statistic of second independent variable = –5.21
t statistic of third independent variable = –6.33
Standard error = $3.05 million
Coefficient of variation = 113.54 percent
Number of observations = 165 lots, 13 programs

The third CER has coefficients that indicate a unit learning-curve slope of 78.8 percent and a production rate efficiency slope of 72.6 percent, and that SE/PM costs per aircraft increases by less than half as air vehicle cost doubles. The CER predicts that SE/PM costs per aircraft declines as more units are produced and production rates increase (up to the maximum yearly quantity), and increases with the AV T100 cost. The CER is based on 160 observations (yearly production lots) of a variety of types of Navy and Air Force aircraft production programs.

The standard error of the estimate is larger than the average SE/PM costs per aircraft in the dataset. Figure 5.9 shows that the CER (represented by the diagonal line) does a poor job of predicting SE/PM production costs when compared with the actual costs (represented by the squares in the figure); however, it does a better job of predicting the overall sample of program costs in the dataset, such that the data is more evenly distributed against the prediction line. As mentioned earlier, the CER should be used only as a rough cross-check, preferably in conjunction with an analogy of the same type of aircraft produced by the same contractor.

The final functional form that we tested was a linear functional form variation using the AV T100 cost and the production quantity rate ratio as the independent variables. This CER gave us better statistical results than the log-log form with the production rate ratio variable, but it does not exhibit a decline in SE/PM costs per aircraft as more aircraft are produced cumulatively. The poorer fit of the log-log

Figure 5.9
Actual Versus Predicted SE/PM Costs for Aircraft Production Programs (FY03 $ per aircraft): CER 3

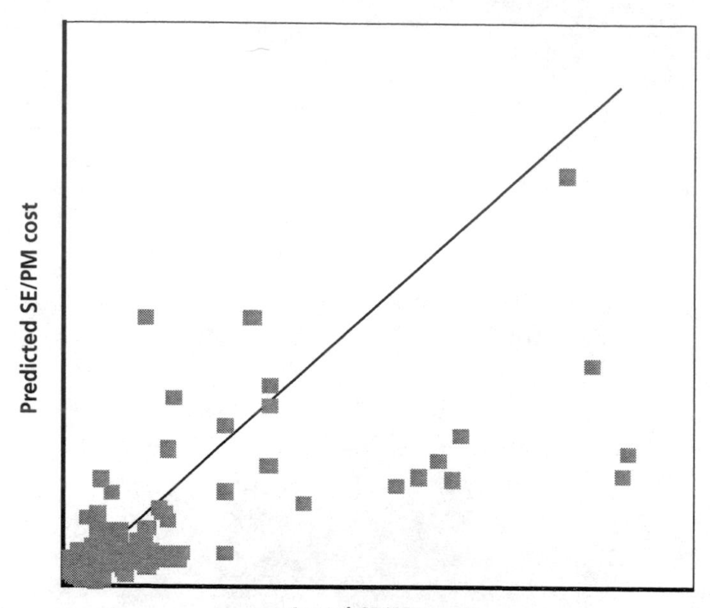

RAND *MG413-5.9*

CER that we developed using a cumulative quantity variable indicates that SE/PM per aircraft in production tends not to decline over time or as more units are produced in a classic learning-curve form, at least after the first few production lots. This fourth CER has coefficients that indicate that SE/PM per aircraft declines as production rates increase, and increases in proportion with the AV T100 cost. The fourth CER and the statistical results are as follows:

Production SEPM Cost per Aircraft (FY03$K) = 1125 + (–2200) * ($Q_n/Q_{max}$) + 0.09108 * AV T100 (FY03$K)
Adj. R^2 = 52.02 percent
F-statistic = 87.18
t-statistic of intercept = 2.27
t-statistic of first independent variable = –2.89

t-statistic of second independent variable = 13.18
Standard error = $2.84 million
Coefficient of variation = 105.57 percent
Number of observations = 165 lots, 13 programs

Figure 5.10 shows how closely the CER predicts SE/PM production costs as compared with the actual costs. As was the case with the prior CERs, there is still quite a large amount of variation between the actual SE/PM costs of the programs and the prediction line. Like the third CER we tested, this CER does a better job of reflecting all of the data in the dataset than does the traditional learning-curve approach attempted in the first two production CERs. Given the poor fit statistics, as with all the aircraft production CERs, this methodology should be used only with extreme caution.

Figure 5.10
Actual Versus Predicted SE/PM Costs for Aircraft Production Programs (FY03 $ per aircraft): CER 4

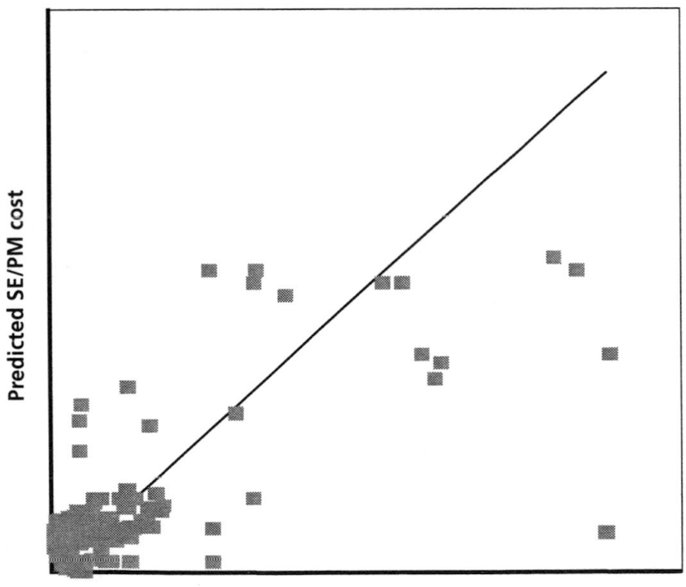

RAND *MG413-5.10*

An important reason for the poor fit of CER equations to the historical data is the wide range of cost improvement slopes observed among historical aircraft production programs. Figure 5.11 summarizes the wide range of cost improvement slopes in our dataset, which was calculated using cumulative quantity only as the independent variable.

Given the above analysis, we tend to prefer the overall better fit provided by the fourth CER. It has the best fit of all the alternatives and does not fall prey to having the problem of poorly representing SE/PM costs for large programs. As with all the production CERs developed in this study, the error remaining is high, and the CER should be used with caution.

Figure 5.11
SE/PM Cost Improvement Slopes on Aircraft Production Programs

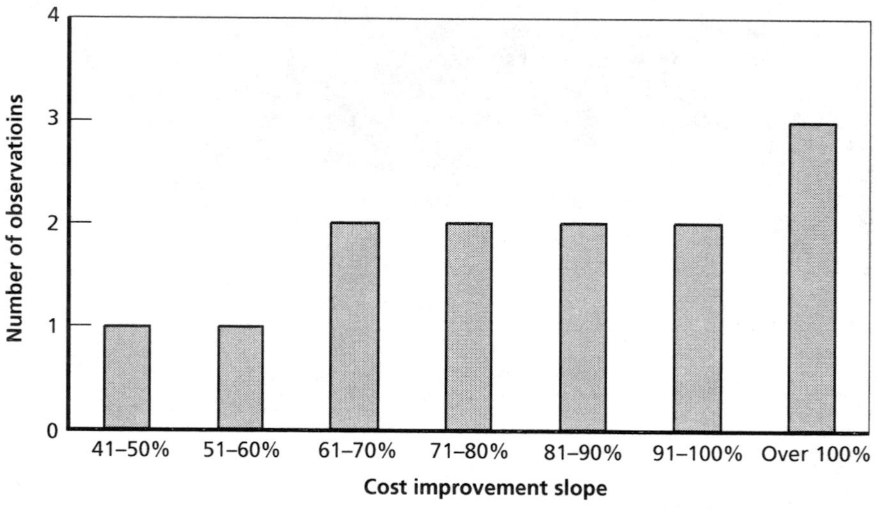

RAND *MG413-5.11*

Guided Weapons Development SE/PM Cost-Estimating Analysis

As with aircraft, we investigated the cost drivers that the contractors we interviewed recommended as a means for estimating guided weapon SE/PM costs. We chose four general categories of variables that account for the following program characteristics: program scope, program duration, physical size, and programmatics. As we did with aircraft programs, we chose quantitative parameters that could be used to represent the general cost drivers for developing CERs.

Guided Weapons Development SE/PM Cost-Estimating Parameters

Most of the parameters that we used in our analyses of weapons development programs were indicators of program size, duration, or complexity. Table 5.6 shows the cost drivers we investigated and the parameters we chose to represent those cost drivers.

The first parameter was a scope variable chosen to indicate the program's size and complexity. The cost of the weapon development program cost minus the cost of SE/PM (TDEVLSEPM) was the parameter we investigated. The next three parameters measure the

Table 5.6
Cost Drivers and Parameters Used for Guided Weapons Development Analysis

Cost Drivers	Parameters for Analysis
Program scope	TDEVLSEPM
Program duration	Months from contract award to first guided launch
	Months from CADT
	Months from contract award to first production delivery (CAFPD)
Physical size	Weight
	Diameter
	Density (weight/cross-sectional area)
Programmatics	Dem/Val or Modification binary or dummy variable
	Contract award year

duration of the development program. The months from contract award to first guided launch (CAFGL) represent the duration of design. The months from contract award to end of development test reflect the duration of total development. Months from contract award to first production delivery (CAFPD) are also meant to reflect the duration of total development. This last parameter is a less satisfactory indicator of the duration of total development in our opinion because it is more influenced by the concurrency of the development and production phases. Nevertheless, we included it in the investigation because it is sometimes more readily available to estimators than the preferable CADT parameter.

The parameters of weight (WT), diameter (DIAM), and density (DEN) (weight/cross sectional area) are physical-sizing parameters meant to scale programs similar to the weight empty variable used for aircraft programs. It was thought that a smaller diameter missile would require denser packaging and greater complexity and would drive SE/PM costs. Also, air-to-air weapons tend to have a smaller diameter and also a smaller weight than air-to-ground weapons. The density variable was used to normalize the weight in a specified package. Table 5.7 shows that air-to-air weapons tend to weigh less, but they have a smaller diameter and cross-sectional area than air-to-ground weapons. The ratio of weight and cross-sectional area provides a useful measure of overall weapon density.

Table 5.7
Summary of Weapons Program Physical Parameters

	Average Weight (pounds)	Average Diameter (inches)	Average Cross-Sectional Area (square inches)	Average Density (Weight/Cross-Sectional Area in pounds per square inch)
Air-to-air weapons	453	7.9	57	9.4
Air-to-ground weapons	1,598	16.7	237	7.1

Two parameters were investigated to see how programmatic drivers affect SE/PM costs. A Dem/Val or modification (DVMOD) program binary or dummy variable was used to compare the larger FSD/EMD programs to these generally more modest efforts. Contract award (CA) year was included in the analysis to allow us to explore the effects of trends over time, but not necessarily for potential inclusion in a CER.

Guided Weapons Development SE/PM Cost-Estimating Relationships

Repeating the process that we followed to investigate estimating parameters for aircraft programs, we used regression analysis to determine the best independent variables to predict SE/PM costs. We used stepwise regression of the candidate independent variables for weapon SE/PM costs, and applied the same four criteria as we did in the selection of independent variables for aircraft SE/PM costs. The variables that we evaluated are shown in Table 5.8, along with the reason for their exclusion from the CER if they were excluded.

Contract award year was not included as a parameter for development programs because the two largest guided weapons development programs in the dataset were the short-range attack missile

Table 5.8
Parameter Analysis Results for Guided Weapons Development Programs

Parameters	Results from Regression Analysis
CA year	Outlier influence
Months from CADT	Not significant
Months from CAFGL	Not significant
Months from CAFPD	Not significant
DEN (weight / cross sectional area)	Wrong sign
DIAM	Not significant
TDEVLSEPM	Included
WT	Not significant
DVMOD binary or dummy	Not significant

(SRAM) and advanced medium-range air-to-air missile (AMRAAM) programs awarded in 1966 and 1981, respectively. These observations are overly influential because of their size, and because they are older programs, they distort analyses of trends.

The preferred CER for estimating guided weapons development SE/PM costs is a linear relationship with TDEVLSEPM. The CER forecasts contractor SE/PM costs (less G&A cost)[14] in FY03 dollars. The CER (in boldface) and summary statistics are as follows:

Guided Weapon Development SE/PM Cost (FY03$K) = 11870 + 0.2263 * TDEVLSEPM (FY03$K)
Adj. R^2 = 83.25 percent
F-statistic = 179.94
t-statistic of first independent variable = 13.41
Standard error = $27.73 million
Coefficient of variation = 51.16 percent
Number of observations = 37

This CER suffers from the same problem as the aircraft development SE/PM CER. The weapons development data set of 37 programs encompasses modification programs totaling a few tens of millions of dollars to full-scale development programs of more than a billion dollars. Although the CER fits the data reasonably well, as shown in Figure 5.12, its standard error is more than 50 percent of the average SE/PM cost in the data set. Analysts who are estimating small development programs such as minor modifications or prototype efforts should be especially cautious in using this CER, and at a minimum should cross-check the results against the costs of an analogous historical program.

---

[14] The average G&A cost for guided weapons system development programs is 12 percent, with a standard deviation of 6 percent.

Figure 5.12
Actual Versus Predicted SE/PM Costs for Guided Weapons Development Programs (FY03 $)

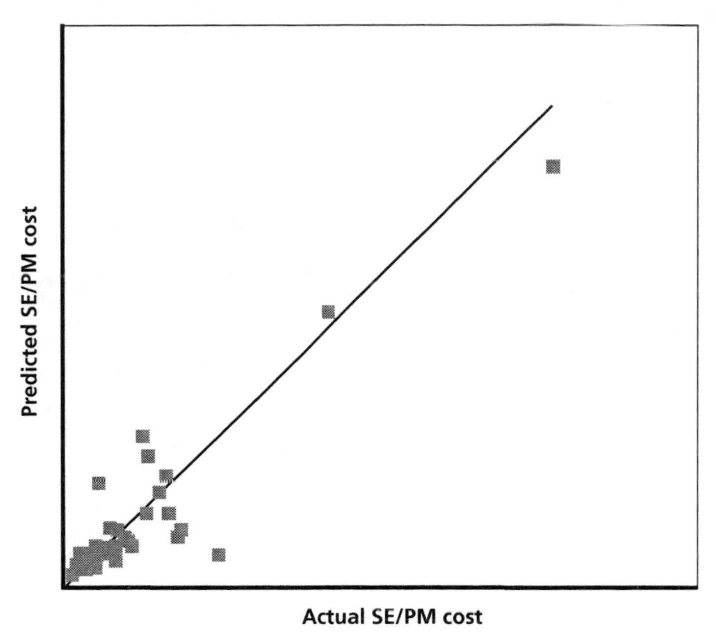

RAND *MG413-5.12*

Guided Weapons Production SE/PM Cost-Estimating Analysis

Using the broad categories of cost drivers recommended by industry personnel as a way to estimate SE/PM costs, we developed a more specific set of parameters. These parameters capture the general nature of the recommended cost drivers in a quantifiable fashion. Using regression analysis, we used these parameters along with historical cost data to generate cost-estimating relationships for guided weapons in production.

Guided Weapons Production SE/PM Cost-Estimating Parameters

Most of the parameters that we used in our analyses of weapons production programs are indicators of product complexity, or were included to explain changes in SE/PM costs from year to year during the production phase. The list of parameters is shown Table 5.9.

Three parameters were chosen to represent the scope of the program. The first two are different versions of measuring recurring unit cost[15] of the weapon. One measures the value at the one-hundredth production unit (Weapon 100th unit cost [WPN T100]) and the other at the one-thousandth unit cost (WPN T1000). The SEPM DEV cost was also seen as a potential indicator of program scope and complexity.

The physical parameters of weight, diameter, and density were expected to be associated with SE/PM costs in production. As we stated earlier, weight has traditionally been used in cost analysis as an

Table 5.9
Variables Used for Guided Weapons Production Analysis

Cost Drivers	Parameters for Analysis
Program scope	WPN T100
	WPN T1000
	SEPM DEV
Physical size	WT
	DIAM
	DEN
Programmatics	LOT MP
	RATE
	Q_n/Q_{max}
	LLOT binary or dummy variable

[15] The guided-weapon recurring unit cost came from Functional Cost Hour Reports (FCHR) for the various programs and production lots. It includes the manufacturing (labor plus purchased equipment), recurring engineering, sustaining tooling, quality control, and other recurring costs. It is separate from nonrecurring, engineering change order, SE/PM, systems test and evaluation (ST&E), training, data, support equipment, operational site activation, fielding, and other procurement costs from the FCHRs. It does not include G&A, cost of money, or profit.

indicator of complexity. We also believed that complexity could be indicated by the amount of miniaturization required in overall packaging of the weapon. We included analysis of the diameter and cross-sectional area of the weapons programs as a means for testing this hypothesis.

The last set of parameters is related to the programmatic aspects of the dataset. We used the variable of lot midpoint (or cumulative quantity) to reflect the expectation that declining SE effort would be required after deficiencies discovered in testing are corrected in early production lots. Thus, we expected that weapon production SE/PM costs would be inversely associated with LOT MP. We expected that this relationship would be smoother and more consistent than we observed for SE/PM costs in aircraft production because guided weapons production programs tend to have fewer major configuration changes. Major configuration changes in weapons programs are usually the result of new development programs and are designated as a new and separate weapon production program. Algebraic LOT MPs were calculated using the average slope of the entire dataset.

We included the variable of rate per year to reflect the idea that recurring SE/PM is a fairly constant level of effort, so we would expect that the amount of effort per weapon would increase as rate per year decreased on a given program. As with our aircraft analysis, we also used the production quantity rate ratio in our analysis.

The last lot binary or dummy parameter was used to reflect the increased SE/PM effort often observed at the end of production programs. Some observers have reasoned that there are "shut down" costs inherent in the last lot. We tried to test the significance of this theory using a binary or dummy variable.

Guided Weapons Production SE/PM Cost-Estimating Relationships

Using regression analysis, we determined the set of variables we would use for developing cost-estimating relationships. Table 5.10 shows the results of the analysis, along with the reason for each parameter's exclusion from the CER if it was excluded.

Table 5.10
Parameter Analysis Results for Guided Weapons Production Programs

Parameters	Results from Regression Analysis
DEN	Not significant
DIAM	Not significant
LOT MP	Included
RATE	Included
Q_n/Q_{max}	Included
SEPM DEV	Not significant
WPN T100	Correlated with other parameter
WPN T1000	Included
WT	Not significant
LLOT binary or dummy	Not significant

As with the analysis to develop CERs for aircraft, we investigated several CERs in the functional form of Y = A * X ^ B * Q ^ R for weapons production programs using the different parameters as independent variables. As mentioned previously, the A value represents the cost of the first unit and provides a way of scaling the equation on the y-axis. The X value is the algebraic midpoint of the lot in question, and the Q value is the quantity procured in the specific lot. The best results were obtained using the following two CERs.

For the first CER, used the above functional form utilizing the recurring cost of the T1000, the lot midpoint, and the yearly production rate as independent variables. The CER yields a SE/PM cost per unit (less G&A costs)[16] in constant FY03 dollars. The regression of the 103 observations resulted in the following equation (in boldface) with the associated statistics.

Guided Weapon Production SE/PM Unit Cost (FY03$K) = 19.42 * WPN T1000(FY03$K)^ 1.035 * LOT MP ^ (–0.2238) * RATE ^ (–0.5831)

[16] G&A costs varied by program and by production lot. The average G&A cost is 8.7 percent, with a standard deviation of 1.7 percent.

Adj. R^2 = 70.51 percent
F-statistic = 501.42
t-statistic of first independent variable = 17.15
t-statistic of second independent variable = –5.25
t-statistic of third independent variable = –7.93
Standard error = $112.31 million
Coefficient of variation = 98.86 percent
Number of observations = 103 lots, 15 programs

As compared with the similar analysis performed on aircraft production data, the guided weapon CER did a much better job of predicting SE/PM costs using the learning and rate functional form. The resulting cumulative quantity curve is 85.6 percent, and the rate curve is 66.8 percent. The recurring unit cost of the weapon is positively associated with SE/PM cost per unit, indicating that more-expensive weapons require more associated SE/PM costs. Nevertheless, the variance explained by the model as indicated by the R^2 is marginal, and the error remaining is high. The CER, like the others, should be used with caution. The plot of the actual costs versus the predicted values is given in Figure 5.13.

As with our analysis for aircraft programs, we wanted to investigate what the results would be using the functional form that utilizes the ratio of yearly lot size to the maximum lot size. We tried two variations using this independent variable. The first version was again a logarithmic function using guided weapons' recurring cost at T1000, lot midpoint, and the quantity ratio Q_n/Q_{max} as the independent variables. The second was a linear-type equation using only WPN T1000 recurring cost and the quantity ratio as independent variables, similar to the methodology used for aircraft production programs.

From the analysis using the logarithmic version of this functional form, we were able to improve the fit statistics of the resulting CER. In this formulation, the cumulative quantity curve is 74.7 percent (steeper than the first CER), and the rate efficiency curve is 87.1 percent (flatter than the first CER). The recurring unit cost of the

Figure 5.13
Actual Versus Predicted SE/PM Costs for Guided Weapons Production Programs (FY03 $ unit cost): CER 1

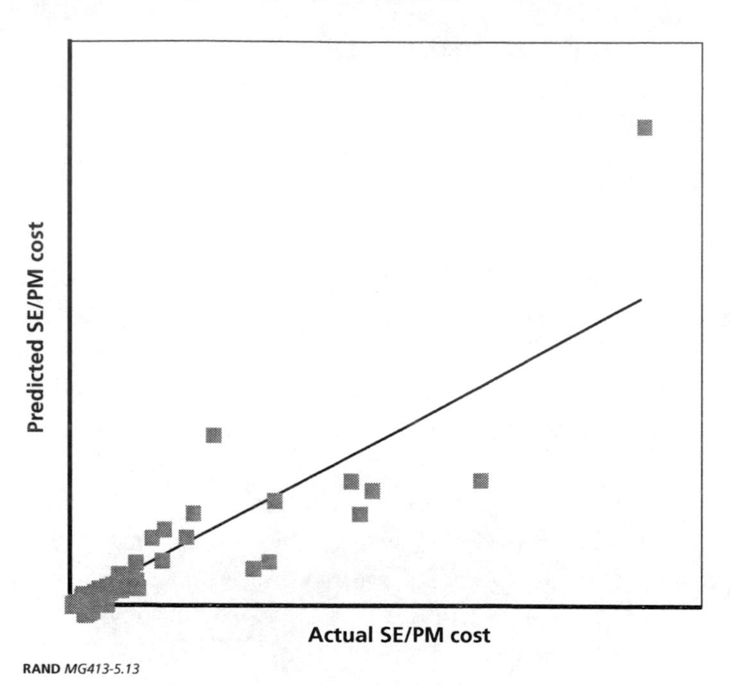

RAND *MG413-5.13*

guided weapon is positively associated with SE/PM costs per weapon. The regression (in boldface) and resulting fit statistics are as follows:

Guided Weapon Production SE/PM Unit Cost (FY03$K) = 0.2879 * WPN T1000 (FY03$K) ^ 1.339 * LOT MP ^ (–0.421) * (Q_n/Q_{max}) ^ (–0.1996)
Adj. R^2 = 83.29 percent
F-statistic = 316.09
t-statistic of first independent variable = 19.52
t-statistic of second independent variable = –9.67
t-statistic of third independent variable = –2.59
Standard error = $84.54 million
Coefficient of variation = 74.42 percent
Number of observations = 103 lots, 15 programs

The second guided weapon production SE/PM CER fits the data better than does the first guided weapon production SE/PM CER. The second CER explains more of the variation in the data and has a smaller standard error than the first CER. The t-statistic for the quantity ratio variable is smaller than the t-statistic for the yearly rate variable in the first CER, but it is significant above the 95 percent confidence level. The plot of actual versus estimated values is shown in Figure 5.14.

The change in magnitude of the cumulative-quantity slopes and rate variables from the first production CER to the second CER is due to the high correlation between these independent variables, a condition known as multi-collinearity. The total cost improvement calculated using the rate and cumulative quantity variables is similar for both CERs. In the first formulation, the regression attributes

Figure 5.14
Actual Versus Predicted SE/PM Costs for Guided Weapons Production Programs (FY03 $ unit cost): CER 2

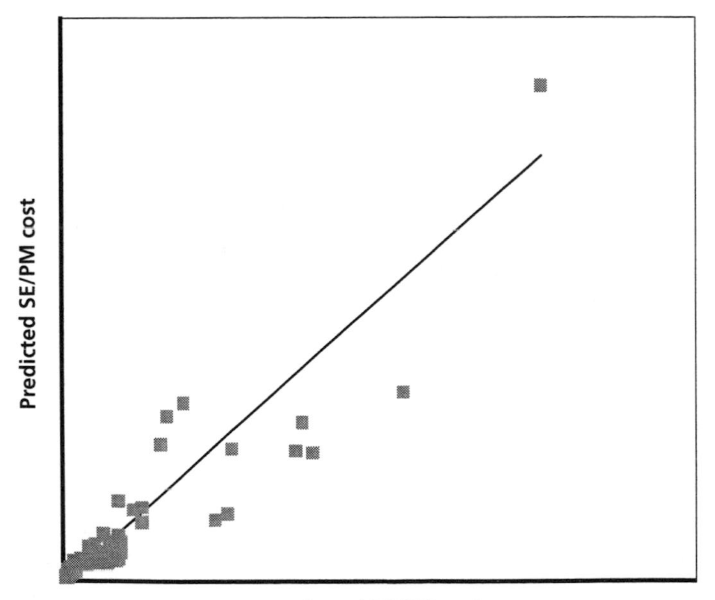

RAND *MG413-5.14*

more cost improvement to the rate variable. In the second formulations, more cost improvement is attributed to cumulative quantity. Multi-collinearity between cumulative quantity and rate variables is a common problem in analyzing production data, and although statistical techniques exist to compensate for this condition, determining the "true" contributions of rate variables and cumulative quantity in these instances is often impossible.

The third formulation we tried for developing a guided weapon production CER uses a combination of the T1000 recurring cost and the quantity ratio as independent variables in a linear-type equation. Because this formula produced the best results for aircraft production programs, we wanted to see how it performed for weapons production programs. Although all of the variables are significant at the 99-percent confidence level, the CER fit the guided weapons data much worse than the logarithmic formulations we developed using lot midpoint and rate variables.

Even though the logarithmic versions of the above CERs fit the data better than the linear version, there is still a large degree of variation in the cost improvement slopes in our dataset. Figure 5.15 shows the range and distribution of slopes calculated by cumulative quantity only (without a rate effect). Note that the distribution of slopes is much tighter than the SE/PM production cost slopes for aircraft, as shown in Figure 5.11.

Summary of Aircraft and Guided Weapons SE/PM Cost-Estimating Relationships

Our analysis to derive CERs for estimating SE/PM costs for aircraft and weapons programs started with determining cost drivers logically associated with SE/PM costs. These cost drivers were largely based on discussions with cost estimators in the government and industry. We next determined quantitative parameters that could be used to represent the suggested cost drivers. Using historical cost data and regression analysis, we developed several CERs that can be used to estimate SE/PM costs.

Figure 5.15
SE/PM Cost Improvement Slopes in Guided Weapons Production Programs

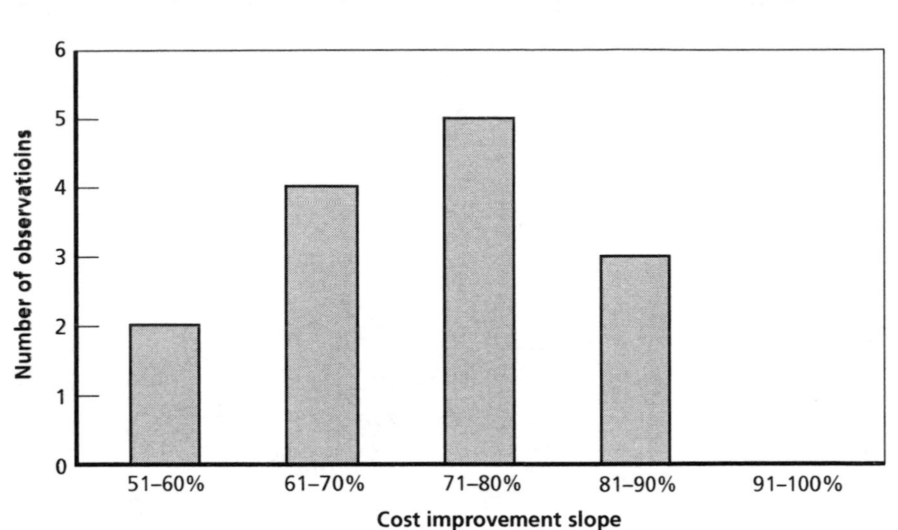

RAND *MG413-5.15*

Given the above analysis, we tend to prefer the overall fit provided by the second CER. As with all the production CERs developed in this study, the error remaining is high, and the CER should be used with caution.

We found that SE/PM costs for aircraft development programs were related to the design duration (as measured by time from contract award to first flight) and to the overall program size (as measured by the development program cost minus SE/PM costs). The resulting CER fit the data rather well, but showed somewhat large variation.

Estimating the SE/PM costs for aircraft in production was difficult due to the highly variable cost data. We investigated to see if certain changes in the production program (such as the end of the initial model of aircraft and the onset of FMS) could be used to explain this variation. Unfortunately, neither of these changes explained the variation in SE/PM production cost. We developed two CERs that used the logarithmic formulation and one CER that used a linear formulation. None of these formulations gave very good statistical results. Of the three, the linear formulation using T100 air vehicle cost and the ratio of the lot quantity to the maximum production lot size provided the best results.

For guided weapons programs, SE/PM development cost was found to be related to the overall development cost of the program (less the cost of SE/PM). The CER fit the data reasonably well, but there was still quite a large amount of variation in the data.

The guided weapon production data showed much more consistency (similar to a traditional learning-curve shape), which allowed us to relate weapon production SE/PM unit costs to the overall weapon unit cost, lot midpoint, and the ratio of the lot quantity to the maximum production lot size. The CER fit the data reasonably well, but as with the rest of the CERs, showed a large variance.

A few words of caution should be expressed regarding use of these CERs for estimating. First, the data used for these relationships showed a wide amount of variability. Second, as mentioned in Chapter Two, there is also some difference in the definition of the cost content among contractors and among programs performed by the same contractor that could increase the variation. As such, cost analysts should also consider the use of other techniques (such as analogies or a bottom-up approach) if more information is known about a program. Finally, another drawback of the CERs is that one of the input variables in each of them is based on an estimate of the other development or production costs. Uncertainty in the SE/PM estimate is increased by uncertainty of the estimates of other program costs as inputs to these CERs.

In the next chapter, we investigate how some of the new acquisition approaches could affect how SE/PM costs are estimated. This information can be used to see if adjustments are required to estimating relationships based upon historical data.

CHAPTER SIX

How New Acquisition Practices Will Affect SE/PM Cost Estimates

One of the objectives of this analysis was to determine if the DoD's recent acquisition-process initiatives—i.e., acquisition reform—had an effect on how SE/PM costs should be estimated for future programs. The three main acquisition initiatives we investigated were the effects of the reduction in the number of MILSPECs and MILSTDs; the use of IPTs; and the use of evolutionary acquisition, which recognizes, at the outset, the need for future improvements to improve capability. Because there are limited cost data for programs that used these new practices and few programs completed post-acquisition reform, determining what quantitative effects these initiatives have on SE/PM costs is especially challenging. It is also difficult to determine if the new acquisition practices or other program characteristics are causing changes in SE/PM costs.

As discussed earlier in this report, our general method for determining differences in SE/PM costs between programs that have used the new acquisition processes and those that have not was to compare the cost data from programs using the new practices to the overall sample of programs in that specific data group. We used univariate analysis[1] to determine if the SE/PM cost for a program using

[1] Univariate analysis provides an analysis of the values for a single variable. It includes measures of central tendency, dispersion, shape, and the predictive capability of the mean as a sample estimator. This approach shows only how the data point of interest differs from the rest of the data. It is not able to determine if the hypothesized cause for the difference is actually the reason for the difference.

the new acquisition approach was significantly different from that of the overall sample.

To examine the effect of the reduction in MILSPECs and MILSTDs, we compared the SE/PM cost from two programs using the new acquisition practices to the rest of the guided-weapons development cost dataset. To examine the effect of the use of IPTs, we looked at the SE/PM cost of two programs that were among the first to use IPTs during their development and compared their SE/PM cost with the SE/PM cost from other historical aircraft programs. For the third new acquisition practice, the use of EA, we took a more qualitative approach to the analysis and investigated cost data from a surrogate program, discussed later in this chapter. Because the use of EA on large aircraft or guided-weapons development contracts is a rather recent occurrence, we were limited in the amount of quantitative analysis we could perform in this area. Our approach was to interview a cost analyst working on the Global Hawk program, which is currently using EA, to see what observations he could provide regarding cost analysis of such a program. We also tried to see what quantitative information could be gleaned from the JSOW program, which concurrently developed three variants of a missile, as a surrogate for an EA program.

Reduction in Military Specifications and Military Standards

One of the acquisition changes DoD instituted in the mid- to late-1990s was a push to reduce the number of MILSPECs and MILSTDs to alleviate what was perceived as overly restrictive and burdensome requirements placed on DoD contractors. The thinking was that significant savings could be realized if some restrictions were eliminated. Two pilot programs—JDAM and JASSM—were used to test how this new acquisition approach worked.

We were interested in seeing if SE/PM costs were affected for a program with fewer MILSPEC and MILSTD restrictions. We tested this hypothesis by comparing the SE/PM costs from the JDAM and

JASSM programs with the cost data in the rest of our sample of guided weapons development programs.[2] Figure 6.1 plots the SE/PM cost distribution for the guided weapons development programs in our dataset. When we compared the SE/PM costs for the JDAM and JASSM programs with the SE/PM costs for the rest of the sample, we found that the costs for the two programs were within one standard deviation of the mean of the SE/PM cost for the sample. We concluded that adjusting SE/PM estimates probably is not necessary for programs that eliminate MILSPECS and MILSTDS.

Figure 6.1
Guided Weapons Development SE/PM Costs for Comparison with Programs with Fewer MILSPEC and MILSTD Restrictions

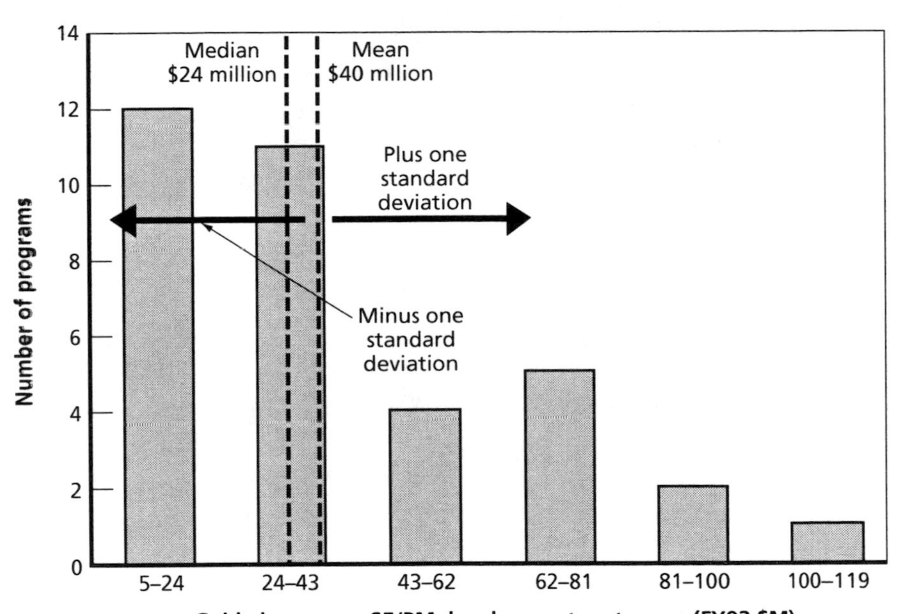

RAND *MG413-6.1*

[2] We used 35 of the 37 historical DoD guided-weapons programs for this analysis. Two programs were eliminated because their SE/PM cost was more than 2.5 standard deviations from the sample median value.

Our interviews with contractors seemed to corroborate this finding. They agreed that the reduction in MILSPECs and MILSTDs decreased the amount of required documentation; however, reducing the use of MILSPECs and MILSTDs did result in additional costs to them. Because commercial practices are to largely replace the military standards, contractors conceded that more effort was required on their part to define an appropriate specification and to make sure that all the subcontractors understood what was required, rather than relying on a government-imposed definition of requirements.

Use of Integrated Product Teams

We also performed a univariate analysis to determine if the use of integrated product teams had any effect on SE/PM costs. Figure 6.2 shows SE/PM cost divided by total program cost less SE/PM (labeled as "SE/PM cost percentage of non-SE/PM cost") for the aircraft development programs in our dataset. We compared the SE/PM cost percentage for the F/A-18E/F and F/A-22 EMD programs to the overall dataset of aircraft development programs. We found that the SE/PM cost percentages for these two programs was not very different from the SE/PM cost percentages of most of the programs in the dataset. One program was within one standard deviation from the mean, and the other program was slightly higher than one standard deviation from the mean.

One thing to keep in mind about these two programs is that they followed vastly different approaches to acquisition.[3] Although the F/A-18E/F program used IPTs in development, it benefited from the use of many previously developed technologies and was performed by the same contractor team that worked on the predecessor F/A-18C/D, reducing the overall complexity of program development. In contrast, the F/A-22 program pursued significantly new

---

[3] For further discussion of the F/A-22 and F/A-18E/F development programs, see Younossi et al., 2005.

Figure 6.2
Aircraft Development SE/PM Cost Percentages for Comparison with Programs Using IPTs

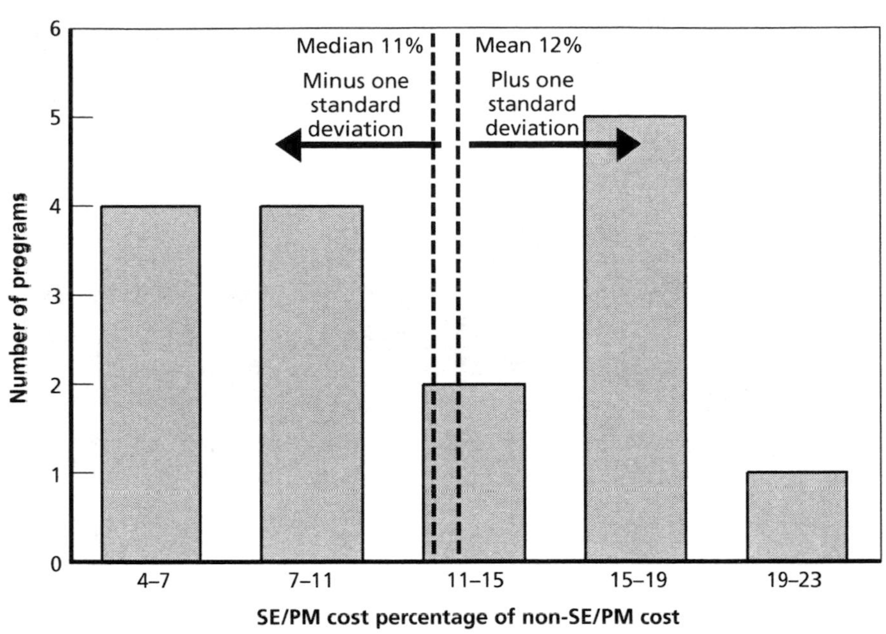

RAND MG413-6.2

performance goals in several technologically challenging areas simultaneously. Also, the F/A-22 required much more time to develop than the F/A-18E/F. If the SE/PM cost element contained level-of-effort tasks, the extension of the schedule in development would also drive up SE/PM costs.

Cost estimators from major prime contractor companies said that they believe the use of IPTs would result in a program costing more than a traditional program.[4] They attribute the additional cost to an increase in SE/PM effort due to additional time in coordinating with the government counterpart in the IPT before proceeding with

[4] In addition, Cook and Graser (2001) quoted a contractor saying that IPTs as a rule add 10 to 20 percent to the initial design cost.

the design process. IPTs, in their opinion, tend to increase the scope of the work that contractors must perform.

Boeing supplied us with a comparison of the engineering hours from the F/A-18A/B development program (a non-IPT program) and the F/A-18E/F program. The comparison showed that the F/A-18E/F program used more engineering hours for SE. However, the overall engineering hours decreased from the A/B to the E/F program. Much of the decrease was due to reduced test-and-evaluation engineering, probably the result of carryover of much of the avionics subsystems from the prior version of the aircraft.

We conclude that it may be necessary to account for the use of IPTs when performing an SE/PM cost estimate for a new program. A cost analyst needs to give careful consideration to the program being used as the basis of the comparison estimate. Choosing a program with a scope and technical complexity that are similar to those of the new program should help to eliminate potential differences between the two programs.

Use of Evolutionary Acquisition

Evolutionary acquisition attempts to speed delivery of a product to the end user. EA can potentially shorten development time in order to capture rapid technological improvements that might be incorporated in the end product. As described in Chapter Three, the process requires additional capability to be released in a series of overlapping "spirals". Each spiral has some dependence on the work from the preceding spiral but also adds unique capability. Each spiral also has its own oversight review process that needs to be addressed.

EA would seem to increase the complexity of the effort required by SE/PM personnel because requirements must be redefined for each successive iteration or spiral. Presumably, this constant change in the development baseline, in which one design will be fielded in the short term while another design is being developed, makes it difficult to maintain configuration control of the different systems resulting from each spiral. This situation results in a continual trade-off between changing needs and desired capability and the finite resources to perform a task.

However, there is some logic in SE/PM programs' use of EA leading to cost savings. As explained in Chapter Three, EA attempts to incrementally deliver capability with the understanding that future technological improvements will occur. If each increment represents delivery of a mature technology, the process may reduce the complexity of the SE/PM tasks within a program.

Because EA is a relatively new approach to major weapons-systems development programs, the amount of quantification we could perform in this area was limited. We used a combination of three approaches for analyzing the potential effects on SE/PM cost estimating due to the use of EA. First, we had discussions with a cost analyst working on the Global Hawk program on how he took EA into account when developing program estimates. Second, we asked cost analysts working for prime contractors what they believe are the implications of EA for SE/PM costs. Finally, we investigated SE/PM costs from the JSOW program as a surrogate for a true EA program to see if we could glean any quantitative results.

The Global Hawk cost analyst provided the following insights and lessons learned from the program. Program engineers were able to rapidly develop the basic air vehicle and achieve first flight within 38 months of contract award. However, the rapid rate of development made it difficult for the regulatory and oversight groups to keep up with the program.[5] The program managers were required to seek approval for each spiral every time significant work content was changed. Also complicating the situation was the division of each spiral into several smaller efforts (referred to as "increments"). Each increment had to be modular and estimated separately to allow an increment to be delayed to future spirals if it encountered budgetary, technical, or scheduling difficulties.

---

[5] The author of a briefing on the Global Hawk program said that, as of the date of the briefing, the program had "experienced the equivalent workload of three milestone reviews, plus an independent full-scale engineering review all within the past 22 months" (Pingel, 2003).

Another complication mentioned by the Global Hawk cost analyst is that spiral development runs counter to the sequence of development, production, and operating and support lifecycle phases. Successive spirals being released to the end user result in overlap of the lifecycle phases. A change in design could affect the building of units that are currently in production or could require retrofits for units that are already fielded. These concurrent phases require a large amount of coordination at all times to ensure that the program strategy is continuously followed between the program phases.

The implications of the complexities of EA for cost estimating are that while EA possibly reduces technical risk by allowing for partial solutions to be fielded more quickly, it may incur increased costs due to increased coordination, integration, and logistics activities for SE/PM. Estimators should divide their SE/PM cost estimates into two parts: the SE/PM cost related to specific spirals or increments and the "overlay" SE/PM cost for the effort that continues across multiple spirals and that provides consistency of overall program direction. Because each spiral or increment may be shifted in the schedule according to the priorities of the program's end users, estimators must be able to determine the unique amount of SE/PM cost required for each spiral or increment. This ability allows for rapid, modular changes to be made to the cost estimate. Also, logistics planners prefer that design changes be "settled out" before performing a full LSA. However, the rush to field systems may push a system to the user before a logistics infrastructure has been established to support the fielded units. Also, as various spirals are fielded, they may or may not include retrofitting to bring older units in line with a common configuration, making configuration control difficult to maintain.

Our discussions with contractor cost analysts indicated that EA would require additional SE/PM effort. The contractors also believe that EA is similar to the P3I concept previously used on programs. Given this assumption, we investigated the JSOW program as a surrogate to a true EA program to see if any quantitative information could be drawn from the JSOW program data.

The JSOW program developed three guided-weapons variants based on a common "truck" airframe design. The initial JSOW A version was to carry submunitions for soft targets, while the JSOW B was to carry fewer but larger submunitions for targets such as tanks. The JSOW C version used a unitary warhead for hard, fixed targets. The JSOW C version also added a seeker for more precise terminal guidance. Figure 6.3 shows the overlapping development schedules for the three JSOW variants.

We were able to collect historical cost information from the Naval Air Systems Command cost department on the JSOW program. One difficulty we encountered was that the JSOW A and B variants were developed on the same dispenser program contract, and the JSOW C was let on a separate contract. As such, the SE/PM costs for the JSOW A and B development program are combined in much of the official cost data reports from the contractor. Nevertheless, the Naval Air Systems Command provided the breakout of the common WBS elements, such as SE/PM, for each variant so that all three could be compared.

We discovered from this information that all variants of the JSOW program experienced a higher SE/PM cost percentage than the average SE/PM cost percentage of other guided weapons development programs. Also, the JSOW A variant experienced a much higher SE/PM cost percentage than the two subsequent variants. This

Figure 6.3
Overlapping Program Development Schedules of the Three JSOW Variants

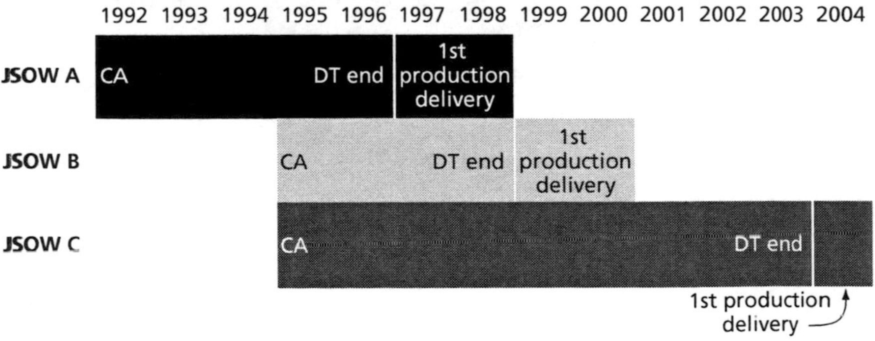

RAND MG413-6.3

may be indicative of the SE/PM effort that is associated with the JSOW A's also providing the core SE/PM functions for the two later variants. Unfortunately, we were not able to obtain a more detailed breakout of the SE/PM costs by variant to clearly show how much SE/PM is devoted to common tasks across all variants and to tasks that apply solely to the specific variants.

Summary

Our investigation into the effects on SE/PM costs from new acquisition initiatives produced mixed results. First, the reduction in military specifications and military standards does not seem to warrant adjustments to standard estimating methodologies employing historical cost information from programs that used MILSPECs and MILSTDs. For programs utilizing IPTs, the results are inconclusive; one program showed no change in SE/PM cost and the other program showed an increase above the average cost. Because evolutionary acquisition is still in its infancy, it was difficult to quantify its potential effect on future SE/PM costs. Using a recent program as a surrogate for evolutionary acquisition, we found an apparent increase in SE/PM costs for multiple variants of a weapon system being developed simultaneously. Cost estimators should also consider whether two types of SE/PM effort should be estimated for programs utilizing EA—one associated with each capability increment and another associated with the overlay SE/PM effort for the program.

CHAPTER SEVEN

Conclusions

SE/PM is an important part of the total cost of developing and producing aircraft and guided weapons systems. While SE/PM costs are not directly attributable to the specific work of designing or producing the hardware or software content of an aircraft and weapons systems program, they represent a significant portion of the "below-the-line" costs for program development and production.

Understanding the process of systems engineering and program management provides a sense of the iterative development steps that match customer needs with the final design and production output. Several tools (such as design reviews) are used by a systems engineer and program manager to monitor and adjust a design, while balancing the necessary adjustments against the technological constraints they pose and the costs they may incur.

SE/PM Cost Trends

SE/PM represents about 12 percent of the total development costs for aircraft programs and about 28 percent of the overall development costs for guided weapons programs. Aircraft development programs have experienced rising SE/PM costs over the past four decades, while SE/PM costs for guided weapons programs have stayed relatively constant. There is an even split between SE and PM costs for aircraft programs, and guided weapons have roughly a 60/40 split between SE and PM costs.

SE/PM costs in production show a large degree of variation from one program to another and within a program from one production lot to the next. Throughout an aircraft program's production run, SE/PM costs are constantly fluctuating. Although SE/PM costs for guided weapons programs also exhibit much volatility, they tend to more closely follow the traditional cost-improvement curve (also known as a "learning curve"; see Chapter Five).

Under the umbrella of acquisition reform, several changes to the acquisition process may affect SE/PM costs. We focused our effort on the following three changes: the reduction in the number of MILSPECS and MILSTDS, the use of IPTs, and the use of evolutionary acquisition.

SE/PM Definitions and Current Cost-Estimating Approaches

One challenge in estimating SE/PM costs is determining what effort is contained in the SE/PM cost element. MIL-HDBK-881 spells out the DoD definition for SE/PM, but there is ambiguity in the definition that can make it difficult to implement consistently in cost-data collection and, as a result, the inconsistency could affect SE/PM cost-estimating methods. The major weapons systems contractors we interviewed for this study had different ways of accounting for these costs, even among programs within the same company. Detailed information from a few programs, however, indicates that many of the major categories of SE/PM costs were consistent among the contractors.

Government and industry use different techniques to estimate SE/PM costs. Government's current estimating techniques aggregate SE/PM costs with the rest of the engineering design costs. Depending on the type of estimate required, contractors use a combination of top-down and bottom-up estimating approaches, including factoring SE/PM costs against the engineering design hours.

New SE/PM Cost-Estimating Approaches

Our goal was to develop a way to directly estimate SE/PM costs using parameters that are specifically related to those costs. By directly estimating SE/PM costs, we hoped to better refine cost estimates and provide a means for isolating any adjustments that are required to account for new acquisition methods. We also wanted to provide a means for time-phasing SE/PM expenditures across a multiyear development program.

We chose to use parameters that are quantitative and readily available early in a program's life cycle. Through interviews with DoD government and industry cost analysts, we identified four general factors that drive SE/PM costs: program scope, program duration, physical characteristics, and programmatic considerations.

Our next step was to determine which quantitative parameters could be used to represent the SE/PM cost drivers and to develop, using regression analysis, cost-estimating relationships. The SE/PM CERs cover four areas of cost: aircraft development programs, aircraft production programs, guided weapons development programs, and guided weapons production programs.

For aircraft development programs, the best CER we could generate is a log-linear function that relates SE/PM costs to the rest of the development cost of a program and the program's design duration (months from contract award to first flight).[1]

Aircraft production programs presented a difficult challenge in developing accurate CERs for estimating SE/PM costs. We investigated the historical cost data to see if the variability was related to changes in aircraft series or to the introduction of foreign military sales, but we were unable to relate these two possible causes to variations in the cost data. We developed three CERs to forecast SE/PM production costs, but all showed poor fit statistics to the historical

---

[1] We further found that SE/PM aircraft development costs are expended in the following fashion: On average, the first third of the cost is expended between the contract award date and critical design review, the second third is spent between CDR and first flight, and the remaining third is spent between first flight and the end of development testing. More details on this analysis are in Appendix F.

data. The SE/PM cost in production showed wide variation in cost improvement when the traditional learning-curve analysis is applied. Due to this uncertainty in the forecasts, caution should be exercised when using the CERs we developed.

Similar to our findings on SE/PM costs for aircraft development programs, SE/PM costs for guided weapons development are related to the total cost of a program (less SE/PM costs). The best functional form was a linear one that fit the data rather well, but showed a large degree of variation.

Because the historical data for guided weapons production programs, our final area of parametric analysis, more closely resembled the typical learning curve, the best CER we were able to develop was a log-linear form that uses the T1000 cost, the lot midpoint, and the ratio of the lot quantity to the maximum lot size as three independent parameters. As was the case with the CER for development cost, the production CER fit the data, but showed a large amount of variation.

To summarize, even though we were able to develop parametric CERs for directly estimating SE/PM costs, we advise cost analysts to compare various techniques and use the one that best suits the particular estimating situation. The data showed some large variations in cost that were not captured by our CERs. Table 7.1 lists various estimating techniques and the advantages and disadvantages of each.

SE/PM development costs can be spread across the development years of a program using either the Weibull distribution (see Appendix F for more information) or a percentage of expenditures at specific milestones.

New Acquisition Practices Have Mixed Effects on SE/PM Costs

When we investigated the effects of acquisition reform on SE/PM costs, we focused on the following practices: reduction in the number of military specifications and standards, the use of integrated product teams, and the use of evolutionary acquisition. Because these practices

Table 7.1
Cost-Estimating Methods and Their Advantages and Disadvantages

Methods	Advantages	Disadvantages
Total engineering CER	Includes all effort Simple to use	Does not identify amount for SE/PM except through allocation May not directly relate to SE/PM cost drivers
Level of effort (bottom-up)	Better visibility of costs at a detailed level Tailored to be program specific	Government cost estimators do not have sufficiently detailed data Use of expert judgment to estimate level of effort does not utilize typical industry average experience Content of detailed SE/PM costs may be missed by a cost analyst using this approach
Percentage of design effort	Can be tailored to specifics of program	Possible vagueness in the DoD definition of the split between design and SE/PM costs
RAND SE/PM CERs	Takes into account experience from many programs Estimates SE/PM costs directly using parameters that are readily available and that are related to SE/PM effort	Assumes consistent definition of SE/PM costs across contractors and programs Industry average may not account for specific program nuances Does not provide details on costs below SE/PM cost element CERs showed a high degree of variation, especially for production programs

are relatively new, there were limited data available to determine the quantitative effects on SE/PM costs. It was also difficult to determine causality between the new practices and SE/PM costs. Overall, we determined, from what data we had, that these new practices have mixed results in changing SE/PM cost estimates from what they would have been based on historical methods.

Using univariate analysis, we determined that programs with reduced usage of military standards and specifications showed no apparent savings in SE/PM costs as compared with other programs in our dataset. We compared the SE/PM cost of the JDAM and JASSM development programs to the SE/PM cost of other guided weapons programs and found that the JDAM and JASSM costs were within one standard deviation of the average for the guided weapons programs in our dataset.

A similar analysis for programs using integrated product teams provided inconclusive results. Data from the F/A-18E/F and F/A-22 development programs that were compared with data from other aircraft development programs showed that SE/PM costs for the two programs were not much different from the overall average SE/PM costs for recent historical programs. Interviews with industry personnel indicated that there may be increased SE/PM effort with IPTs due to greater involvement between the contractor and the government. A comparison of detailed engineering hours for the F/A-18A/B and F/A-18E/F development programs showed an increase in SE engineering labor for the later program. When accounting for the effect of IPTs in new programs, cost estimators should carefully choose the program or programs used as the historical basis for estimates and consider how those programs compare with the new IPT program.

Because programs done under evolutionary acquisition are still in their infancy, we had limited data to determine the effects of a true EA program on SE/PM cost. We performed a quantitative and a qualitative analysis of the possible effects of evolutionary acquisition on SE/PM costs. For the former, we analyzed cost data from the JSOW program that utilized concurrent development of three weapons variants as a surrogate for an EA program. Cost data from this program showed that all variants experience higher-than-average SE/PM costs, with a significantly higher-than-average amount of SE/PM required for the development of the first variant. For our qualitative analysis, we interviewed a cost analyst on the Global Hawk program, one of the first major programs to implement EA. Based on this discussion, we recommend that cost analysts should consider estimating SE/PM costs in two parts. One part of the cost would be directly related to the capability increment or spiral being developed and another part would come from the core SE/PM activities covering the entire program. Interviews with cost-estimating personnel in industry also indicated that additional SE/PM effort is required on

EA programs. As more programs use this acquisition practice, more data should be available to enable better evaluation of EA's impact on SE/PM costs.

Given the large amount of resources required to perform SE/PM activities on military systems, it would be wise to continue conducting research in this area, especially given the desire of many traditional prime contractors to become lead integrators of several complex weapons systems that are networked together. Two examples of this trend are the National Missile Defense System and the Future Combat System, for which a contractor is selected to integrate several programs that are each large in their own right. Also, as is the case with evolutionary acquisition, the systems engineering and program management function may need to address new challenges that could prove to be SE/PM cost drivers.

APPENDIX A

Relationship of Systems Engineering to the Acquisition Life-Cycle Process

The Defense Acquisition Management Framework, as described in Department of Defense Instruction 5000.2 (DoD, 2003b), provides a structured life-cycle process that is to be followed in the acquisition of major programs.[1] The systems engineering process continues throughout all of the life-cycle phases and determines the program's progress through a series of reviews. This appendix provides information on the focus and outcome of those reviews and how they relate to the acquisition life cycle.

As Figure A.1 shows, the Systems Engineering process is linked to the acquisition life cycle through a series of reviews and audits (Defense Acquisition University, 2004a). Milestones A, B, and C in the figure indicate the major decision points at which the program is reviewed for continuation into the next phase of the life cycle. Below the milestones are the life-cycle phases—Concept Refinement, Technology Development, System Development and Demonstration (SD&D), Production and Deployment (P&D), and Operating and Support (O&S). The P&D phase is further subdivided into the LRIP and Full Rate Production (FRP) phases.

---

[1] In October 2003, the Deputy Secretary of Defense canceled the DoD 5000 series guidance documents, and the Joint Chiefs of Staff instituted a new capabilities-based process for identifying current and future gaps in capability. The resulting Joint Capabilities Integration and Development System defined a new process for determining department needs using top-down analyses rather than bottom-up requirement generation.

Figure A.1
Acquisition Life Cycle and Its Links to the Systems Engineering Process

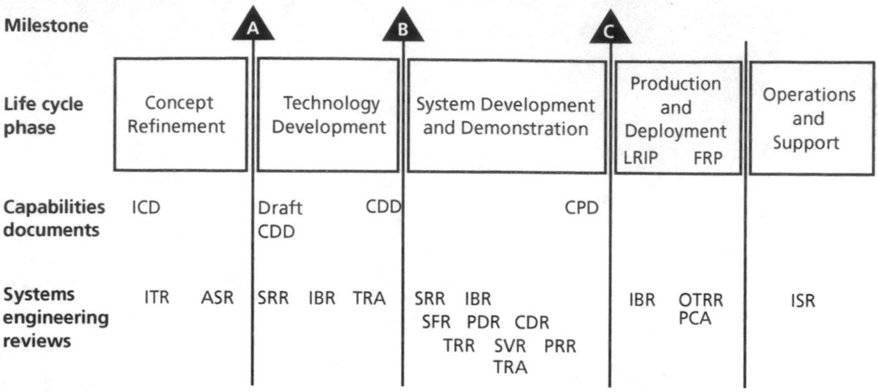

Milestone		A		B		C		
Life cycle phase	Concept Refinement		Technology Development		System Development and Demonstration		Production and Deployment (LRIP, FRP)	Operations and Support
Capabilities documents	ICD		Draft CDD, CDD		CPD			
Systems engineering reviews	ITR, ASR		SRR, IBR, TRA		SRR, IBR, SFR, PDR, CDR, TRR, SVR, PRR, TRA		IBR, OTRR, PCA	ISR

RAND *MG413-A.1*

Below the life-cycle phases is a series of capabilities documents that formally delineate capability needs and solutions to fill capability gaps. The Initial Capabilities Document (ICD) is prepared at the beginning of the Concept Refinement phase and documents the need to resolve a specific capability gap. It further defines the gap in measurable terms and lists material and nonmaterial approaches to addressing the gap. The ICD should be nonsystem specific. It is used to support the decision at Milestone A to enter into the Technology Development phase. The Capabilities Development Document (CDD) is based on the ICD. It outlines the capabilities to be delivered in the first increment of the system and describes the overall strategy for achieving full capability. The CDD provides the authoritative guide for the attributes that are desired upon entering into the SD&D phase. These attributes define performance needs in measurable and testable ways to support the development of the system before the Milestone B decision. The final document, the Capability Production Document (CPD), is used to address production attributes and quantities for procurement. It is further used to support the Milestone C decision to move into the P&D phase. Below the capabilities documents are the Systems Engineering Reviews. We next explain how those reviews relate to the activities being performed during each phase of the life cycle.

During the concept refinement phase, systems engineers perform advanced studies of a broad array of approaches to meet the desired needs of the customer. Two SE reviews occur during this phase: the Initial Technical Review (ITR) and the Alternative Systems Review (ASR). The ITR is designed to assess the capability needs and the approach to meeting those needs and to ensure that the program's technical baseline is sufficiently defined to support a valid cost estimate. The ASR is a comprehensive assessment of the preferred approach to ensure that the resulting set of requirements agrees with the customer's needs. The ASR should be completed before the Milestone A decision is made to enter the technology development phase.

During the technology development phase, systems engineers have the task of converting a required capability into a system performance specification. Three major reviews are conducted during this phase: the System Requirements Review (SRR), the Integrated Baseline Review (IBR), and the Technology Readiness Assessment (TRA). The SRR is used to confirm that progress is being made on ascertaining the technical requirements and that convergence on a balanced and complete solution is being achieved. Most important, this review should determine if all system requirements as stated in the ICD or draft CDD are being met with an acceptable amount of cost, schedule, and technical risk. The IBR establishes a project-performance baseline for measuring earned-value progress during execution of a project. The TRA is a regulatory requirement for assessing the maturity of critical technologies used by the system.

After the Milestone B decision, the program enters the SD&D phase. During this phase, the concept for eventual production and fielding is further developed. Detailed design is performed, and developmental units are produced to establish initial production methods and to support testing.

The SRR and IBR may be repeated at the beginning of the SD&D phase to ensure baselines are well understood by both the government and the contractor. The System Functional Review (SFR) ensures that the system can enter preliminary design and that it meets the system requirements defined in the CDD. The review ensures that the system functional requirements are captured in the system specifications and fully decomposed into lower-level subsystems. The functional baseline is the document that describes the system characteristics and how achievement of these characteristics will be verified.

The Preliminary Design Review (PDR) determines if the design can enter the detailed design phase. It assesses whether the performance specifications of each configuration item are fully captured and ensures that each system function has been allocated to hardware or software elements. The PDR establishes the system-allocated baseline. The allocated baseline documents the configuration items making up the system and allocates the system-level performance requirements to those items. The Critical Design Review (CDR) at this point further evaluates the completeness of the design and the interfaces between each configuration item to ensure that the system can proceed into system fabrication, demonstration, and testing. The functional and allocated baselines again are reviewed and any necessary changes are incorporated into them. The CDR checks for completeness and to ensure that initial builds of the hardware and coding of the software may begin. A final check is made against the requirements to verify the baseline design.

The final major reviews performed under SD&D are the Test Readiness Review (TRR), the System Verification Review (SVR), and the Production Readiness Review (PRR). The TRR is designed to assess test readiness and review the test plans to ensure that the planned test requirements track with the user's needs. The SVR ensures the system is ready to enter LRIP by verifying final product performance and providing inputs to the CPD. The PRR assesses whether the design is ready for production and the contractor has performed adequate manufacturing planning. A second TRA may also be conducted during this phase prior to the Milestone C decision.

Systems engineering during the P&D phase of the life cycle looks for problems that may call for improvement or redesign of the system. An additional IBR may be conducted to once again ensure a performance baseline for measuring earned value. The Operational Test Readiness Review (OTRR) is similar to the TRR, but it is conducted prior to entering formal OT&E. It ensures that the system can proceed into the operational test phase with a high probability of success. The objective is to test the system for suitability and effectiveness for service introduction. The decision to proceed to full-rate production may hinge upon a successful operational test. The Physical Configuration Audit (PCA) is performed to ensure that the test unit representing a production item fully reflects the design baseline specified in the contract. The PCA is usually conducted when the government intends to control the detailed design by acquiring the Technical Data Package. If the government does not plan to do so, the contractor should perform the PCA internally.

Finally, the O&S phase is focused on sustainment of the system as it is fielded. In this phase, new threats are discovered, missions are changed, and deficiencies are brought to light. New technologies and upgrades augment the original concept to meet the new requirements. SE weighs these potential product improvements against the evolving baseline to continually meet the customer's requirements. In Service Reviews (ISRs) may be done to determine the technical status and condition of the fielded system. Large changes to a system may lead to a new program that requires the systems engineering process (complete with formal baselines and reviews) to be repeated.

APPENDIX B

MIL-HDBK-881 Excerpt: Definitions of SE/PM for Cost Reporting

H.3.2 Systems Engineering/Program Management

The systems engineering and technical control as well as the business management of particular systems and programs. Systems engineering/program management elements to be reported and their levels will be specified by the requiring activity.

Includes:

the overall planning, directing, and controlling of the definition, development, and production of a system or program including supportability and acquisition logistics, e.g., maintenance support, facilities, personnel, training, testing, and activation of a system.

Excludes:

systems engineering/program management effort that can be associated specifically with the equipment (hardware/software) element.

Systems Engineering

The technical and management efforts of directing and controlling a totally integrated engineering effort of a system or program.

Includes but not limited to:

- effort to define the system and the integrated planning and control of the technical program efforts of design engineering, specialty engineering, production engineering, and integrated test planning.

- effort to transform an operational need or statement of deficiency into a description of system requirements and a preferred system configuration.
- technical planning and control effort for planning, monitoring, measuring, evaluating, directing, and replanning the management of the technical program.
- (all programs, where applicable) value engineering, configuration management, human factors, maintainability, reliability, survivability/vulnerability, system safety, environmental protection, standardization, system analysis, logistic support analysis, etc.
- (for ships) the Extended Ship Work Breakdown Structure (ESWBS), Configuration Management (811), Human Factors (892), Standardization (893), Value Engineering (894), and Reliability and Maintainability (895) elements.

Excludes:

actual design engineering and the production engineering directly related to the WBS element with which it is associated.

Examples of systems engineering efforts are:

1. System definition, overall system design, design integrity analysis, system optimization, system/cost effectiveness analysis, and intrasystem and inter-system compatibility assurance, etc.; the integration and balancing of reliability, maintainability, producibility, safety, human health, environmental protection, and survivability; security requirements, configuration management and configuration control; quality assurance program, value engineering, preparation of equipment and component performance specifications, design of test and demonstration plans; determination of software development or software test facility/environment requirements.
2. Preparation of the Systems Engineering Management Plan (SEMP), specification tree, program risk analysis, system planning, decision control process, technical performance measure-

ment, technical reviews, subcontractor and vendor reviews, work authorization, and technical documentation control.

3. Reliability engineering—the engineering process and series of tasks required to examine the probability of a device or system performing its mission adequately for the period of time intended under the operating conditions expected to be encountered.
4. Maintainability engineering—the engineering process and series of tasks required to measure the ability of an item or system to be retained in or restored to a specified condition of readiness, skill levels, etc., using prescribed procedures and resources at specific levels of maintenance and repair.
5. Human factors engineering—the engineering process and the series of tasks required to define, as a comprehensive technical and engineering effort, the integration of doctrine, manpower, and personnel integration, materiel development, operational effectiveness, human characteristics, skill capabilities, training, manning implications, and other related elements into a comprehensive effort.
6. Supportability analyses—an integral part of the systems engineering process beginning at program initiation and continuing throughout program development. Supportability analyses form the basis for related design requirements included in the system specification and for subsequent decisions concerning how to most cost effectively support the system over its entire life cycle. Programs allow contractors the maximum flexibility in proposing the most appropriate supportability analyses.

Program Management

The business and administrative planning, organizing, directing, coordinating, controlling, and approval actions designated to accomplish overall program objectives [that] are not associated with specific hardware elements and are not included in systems engineering.

Includes, for example:

- cost, schedule, performance measurement management, warranty administration, contract management, data management, vendor liaison, subcontract management, etc.
- support element management, defined as the logistics tasks management effort and technical control, and the business management of the support elements. The logistics management function encompasses the support evaluation and supportability assurance required to produce an affordable and supportable defense materiel system.
- planning and management of all the functions of logistics.

  Examples are:

  - maintenance support planning and support facilities planning; other support requirements determination; support equipment; supply support; packaging, handling, storage, and transportation; provisioning requirements determination and planning; training system requirements determination; computer resource determination; organizational, intermediate, and depot maintenance determination management; and data management.
  - (for ships) the Extended Ship Work Breakdown Structure (ESWBS), Project Management (897); Data Management (896); and Supply Support (853) elements.

APPENDIX C

Contractor Questionnaire

RAND Systems Engineering and Program Management Cost Estimating Methodologies Project

Background: As part of ongoing research activities under Project AIR FORCE, RAND has been tasked to investigate approaches to develop cost estimates for systems engineering (SE) and program management (PM) for military aircraft and guided weapons programs. We are interested in investigating the nature of SE and PM costs, collecting historical cost data, and developing estimating methodologies that can be used to better estimate SE and PM costs for development and production programs.

1. **Definitions and Breakout:** Government contract cost reporting uses MIL-HDBK-881 to define the content of cost categories including Systems Engineering (SE) and Program Management (PM). Do you agree with this definition? If not, what are the differences in definitions used by your company? Do you combine the costs for these functions, or do you estimate and track them separately? If possible, please provide a more detailed breakout of these costs showing internal cost accounting categories under SE and PM. Are all SE and PM functions charged as direct cost, or are there any indirect costs that would be picked up in the overhead accounts?
2. **Cost Drivers:** What do you believe are the cost drivers for both SE and PM (e.g., size of contract, complexity of program, teaming

with other contractors, etc.)? If program complexity is considered, how is it measured (i.e., number of drawings, weight, SLOC counts)? What role does the duration of the contract play in the estimate of SE and PM costs? How do major milestones (i.e., time to first flight, time to first avionics flight, time to first guided launch) affect SE/PM cost estimates? Does the number of reviews affect SE and PM costs? Does the amount of subcontracted work affect SE and PM costs?

3. **Methodology**: When developing a proposal estimate for SE and PM costs, what approach is generally used? What independent variables are used to estimate the overall SE and PM costs of a typical program? What adjustments need to be considered if using an analogy to a previous program? If possible, please provide an example of a recent estimate on a major program.
4. **Phasing:** During the development phase of a program, what do the SE and PM expenditure profiles look like? For SE, is there a steep ramp up and ramp down leading to a low level-of-effort for the duration of the development? Is the peak spending rate related to a key development milestone, such as Critical Design Review (CDR), first flight, or first guided launch? Is PM essentially a level-of-effort throughout the development? What activities account for this profile and how do they change through the development program? Are SE and PM costs related to the DoD milestone review process (Preliminary Design Review, CDR, Defense Acquisition Board, etc.)? Please provide historical quantitative information to support this profile. Also, if possible, please provide a program schedule that can be used to overlay on the profile to provide the context of the program.
5. **Transition from Development into Production:** How do the tasks involved with SE and PM change when a program moves from the Development phase into the Production phase (sustaining engineering)? Does the SE staff from development continue with the program into production, or are they replaced and reassigned to new development efforts? How do you estimate SE and PM costs in production (percentage of manufacturing, cost improvement curve, fixed headcount)? Are there fixed and variable

aspects to SE and PM costs in production? What role does the number of units produced in a lot (rate) have on SE and PM costs? How does the number of Engineering Change Orders (ECOs) affect SE and PM costs? How do you estimate the amount of SE and PM effort to support ECOs? Is there a smaller cost since there is already an engineering staff supporting the main program? Do multiyear contracts affect SE and PM costs at either the prime or subcontract level?

6. **Trends over Time:** What has been the trend over time across different programs on SE and PM costs? What has been the effect on SE and PM costs of pushing more of the design/trade-off analysis to suppliers? Has the concept of having preferred suppliers changed the amount of SE and PM required at the prime contractor level? How do you estimate/verify subcontract/supplier SE and PM charges? Have new tools played a role in improving the productivity of SE and PM? Do these new tools allow for more design iterations/trade-offs to be explored in the same amount of time at the same cost as historical programs, or is there an actual reduction of effort required?
7. **New Initiatives:** How will the following acquisition initiatives affect SE and PM costs?

   - Commercial Off-the-Shelf (COTS) hardware and software
   - Total Systems Program Responsibility (TSPR)
   - Decreased used of MILSPECS and increased use of performance-based specs
   - Integrated Product Teams (IPTs)
   - Spiral Development and Evolutionary Acquisition
   - Systems-of-Systems concepts
   - What considerations should a cost estimator and program manager have regarding these new initiatives? What work content has changed? If possible, please provide quantitative information to support any estimates for cost savings projections.

8. **Data:** Please provide as much detailed cost data in dollars, work years, and headcounts for SE and PM for both development and

production for all programs at your company. Preferably, we would like to see the efforts phased across program years. Please show headcount comparisons of the SE staff in the development phase as compared with the production phase on specific programs.

APPENDIX D

Variables Used in Developing CERs

Table D.1 lists the independent variables that we chose to represent the potential cost drivers of SE/PM costs in aircraft and guided weapons development and production programs. We used these explanatory variables to develop CERs for estimating SE/PM costs directly. (See Chapter Five for further information.)

Table D.1
Aircraft Variables

Variable	Description/Definition
FF	First flight (FF) date of the aircraft that occurs during the development of the program. For this study, the year that FF took place was used.
CAFF	The duration of the design effort measured by months from contract award to FF date.
CADT	The duration of the development effort measured by months from contract award to end of DT.
NRDEV	NRDEV is the nonrecurring development cost and is normalized to FY03 constant dollars by applying an inflation index at the midpoint of the development program (contract award to end of DT) to the then-year cost reported in a CCDR or similar cost report. It is used as a measure of the size of the development effort. It excludes recurring costs, which are mostly costs from production of test assets.
TDEVLSEPM	TDEVLSEPM is the total development cost less SE/PM and is normalized to FY03 constant dollars by applying an inflation index at the midpoint of the development program (contract award to end of DT) to the then-year cost reported in a CCDR or similar cost report. It is used as a measure of the size of the development effort.

Table D.1—Continued

Variable	Description/Definition
AV/AF	AV/AF is the ratio of development air vehicle cost divided by the airframe development cost. Larger AV/AF values indicate the potential for proportionally more engineering effort not related to a hardware item.
WE	The weight empty parameter is used as an indication of the scope of a program or product.
PROTO	Prototype dummy or binary variable, where a value of 1 indicates a prototype program and a value of 0 indicates all other development programs.
AAD	Airframe already developed dummy or binary variable, where a value of 1 indicates a development program in which the airframe has already been developed and a value of 0 indicates all other development programs.
PVF	Previous version already fielded dummy or binary variable, where a value of 1 indicates a development program in which a previous model of the airframe has already been developed and a value of 0 indicates all other development programs.
AV AUC	AV AUC is the air vehicle average unit cost per lot and is measured in FY03 constant dollars. It is used as a measure of the scope and complexity of a product in production.
AV T100	AV T100 is the air vehicle cost at the one-hundredth unit. It is measured in FY03 constant dollars. The cost is calculated using a cumulative quantity slope only; it does not include the cost of the engine. It is used as a measure of the scope and complexity of a product in production.
SEPM DEV	SE/PM cost in the development program is measured in FY03 constant dollars and includes recurring and nonrecurring costs.
RATE	The yearly production lot quantity for each year of production.
LOT NUM	LOT NUM indicates the production lot of a specific aircraft. The first in a yearly production lot is LOT NUM 1, and later lots are numbered sequentially.
LOT MP	Algebraic lot midpoint calculated using the slope of the cost-improvement curve for SE/PM.
SEPM UC	SEPM UC is the unit cost of SE/PM in production for a given production lot.
SEPM LOT	SEPM LOT is the lot cost of SE/PM in production for a given production lot.
Q_n/Q_{max}	The ratio of the quantity produced in a given lot divided by the maximum yearly lot quantity for the entire production buy.
LLOT	Last lot binary or dummy variable, where 1 indicates the last production lot and 0 indicates all other production lots.

Table D.2
Guided Weapons Variables

Variable	Description/Definition
CA	The year in which the development contract was awarded.
CADT	Duration of the development effort measured by months from contract award to end of DT.
CAFGL	Duration of the design effort measured by months from contract award to first guided launch date.
CAFPD	Duration of the development effort measured in months from contract award to first production delivery.
DEN	The DEN parameter is a ratio of the weight of a weapon divided by the cross-sectional area. It is intended to indicate packaging complexity of a weapon, where weight is measured in pounds and cross-sectional area is calculated as Pi * (diameter in inches^2)/4.
DIAM	Diameter of a weapon measured in inches.
TDEV	TDEV is the total development cost measured in FY03 constant dollars, calculated by applying an inflation index at the midpoint of the development program (contract award to end of DT) to the then-year cost reported in a CCDR or similar cost report. It is a measure of the size of the development effort.
TDEVLSEPM	TDEVLSEPM is the total development cost less SE/PM. It is measured in FY03 constant dollars, calculated by applying an inflation index at the midpoint of the development program (contract award to end of DT) to the then-year cost reported in a CCDR or similar cost report. It is a measure of the size of the development effort.
WT	WT in pounds is an indication of the scope of a program or product.
DVMOD	D&V or modification program dummy (or binary) variable, where a value of 1 indicates a D&V or modification program and a value of 0 indicates all other development programs.
LOT MP	Algebraic lot midpoint calculated using the slope of the cost-improvement curve for SE/PM.
RATE	The yearly production lot quantity for each year of production.
SEPM DEV	SE/PM cost in the development program is measured in FY03 constant dollar and includes recurring and nonrecurring costs.
WPN T100	The recurring cost of the one-hundredth weapon. It is measured in FY03 constant dollars and is calculated using cumulative quantity slope only. It is used as a measure of the scope and complexity of the product in production.
WPN T1000	The recurring cost of the one-thousandth weapon is measured in FY03 constant dollars and is calculated using the cumulative-quantity slope only. It is used as a measure of the scope and complexity of a product in production.

Table D.2—Continued

Variable	Description/Definition
Q_n/Q_{max}	The ratio of the quantity produced in a given lot divided by the maximum yearly lot quantity for the entire production buy.
SEPM UC	SEPM UC is the unit cost of SE/PM in production for a given production lot.
SEPM LOT	SEPM LOT is the lot cost of SE/PM in production for a given production lot.
LOT NUM	The lot number indicating the specific aircraft yearly production lot number, where 1 is the first production lot and lots are numbered sequentially.
LLOT	Last lot binary or dummy variable, where 1 indicates the last production lot and 0 indicates all other production lots.

APPENDIX E

Statistical Correlations for SE/PM CER Variables

The tables on the following pages provide the correlation matrices for the SE/PM CER variables in the datasets used for this study. See Chapter Five for a related discussion.

Table E.1
Correlation Matrix: Aircraft Development

	FF	CAFF	CADT	SEPM DEV	TDEVLSEPM	NRDEV	AV/AF	WE	PROTO	AAD
FF	100%	47%	25%	32%	26%	30%	–15%	16%	1%	–12%
CAFF	47%	100%	83%	87%	81%	85%	–11%	32%	–12%	–18%
CADT	25%	83%	100%	83%	81%	82%	–16%	22%	–12%	–42%
SEPM DEV	32%	87%	83%	100%	95%	96%	–2%	12%	–20%	–17%
TDEVLSEPM	26%	81%	81%	95%	100%	99%	–3%	11%	–29%	–23%
NRDEV	30%	85%	82%	96%	99%	100%	–1%	10%	–28%	–22%
AV/AF	–15%	–11%	–16%	–2%	–3%	–1%	100%	–25%	47%	1%
WE	16%	32%	22%	12%	11%	10%	–25%	100%	36%	–27%
PROTO	1%	–12%	–12%	–20%	–29%	–28%	47%	36%	100%	–22%
AAD	–12%	–18%	–42%	–17%	–23%	–22%	1%	–27%	–22%	100%

Table E.2
Correlation Matrix: Aircraft Production

	LOT NUM	SEPM LOT	SEPM UC	Q_n/Q_{max}	RATE	SEPM DEV	NRDEV	WE	AV AUC	AV T100	LOT MP
LOT NUM	100%	–29%	–25%	–41%	–26%	–27%	–7%	–32%	–31%	–32%	40%
SEPM LOT	–29%	100%	31%	29%	51%	25%	7%	8%	7%	12%	22%
SEPM UC	–25%	31%	100%	–5%	–34%	52%	32%	54%	56%	71%	–28%
Q_n/Q_{max}	–41%	29%	–5%	100%	41%	18%	–11%	9%	–2%	16%	–25%
RATE	–26%	51%	–34%	41%	100%	–11%	–5%	–28%	–26%	–33%	51%
SEPM DEV	–27%	25%	52%	18%	–11%	100%	82%	63%	67%	82%	–17%
NRDEV	–7%	7%	32%	–11%	–5%	82%	100%	54%	58%	64%	1%
WE	–32%	8%	54%	9%	–28%	63%	54%	100%	85%	86%	–30%
AV AUC	–31%	7%	56%	–2%	–26%	67%	58%	85%	100%	81%	–28%
AV T100	–32%	12%	71%	16%	–33%	82%	64%	86%	81%	100%	–36%
LOT MP	40%	22%	–28%	–25%	51%	–17%	1%	-30%	–28%	–36%	100%

Table E.3
Correlation Matrix: Guided Weapons Development

	CAFGL	CADT	CAFPD	WT	DIAM	DEN	TDEV	TDEVLS EPM	SEPM DEV	DVMOD	CA
CAFGL	100%	53%	62%	4%	6%	–5%	23%	25%	12%	–8%	4%
CADT	53%	100%	79%	–14%	–9%	2%	9%	10%	1%	2%	5%
CAFPD	62%	79%	100%	-9%	3%	–14%	5%	6%	2%	–25%	–26%
WT	4%	–14%	-9%	100%	89%	–11%	20%	20%	15%	–43%	–6%
DIAM	6%	–9%	3%	89%	100%	–51%	12%	12%	12%	–41%	–15%
DEN	–5%	2%	–14%	–11%	–51%	100%	4%	6%	–5%	3%	22%
TDEV	23%	9%	5%	20%	12%	4%	100%	100%	94%	–28%	–26%
TDEVLSEPM	25%	10%	6%	20%	12%	6%	100%	100%	91%	–27%	–22%
SEPM DEV	12%	1%	2%	15%	12%	–5%	94%	91%	100%	–31%	–40%
DVMOD	–8%	2%	–25%	–43%	–41%	3%	–28%	–27%	–31%	100%	27%
CA	4%	5%	–26%	–6%	–15%	22%	-26%	–22%	–40%	27%	100%

Table E.4
Correlation Matrix: Guided Weapons Production

	LOT NUM	RATE	SEPM UC	SEPM LOT	LOT MP	SEPM DEV	WT	DIAM	DEN	LLOT	WPN T100	WPN T1000	Q_n/Q_{max}
LOT NUM	100%	14%	–35%	–1%	65%	–8%	–7%	–8%	17%	32%	4%	18%	38%
RATE	14%	100%	–36%	–31%	55%	–16%	–27%	–21%	–8%	23%	–46%	–49%	37%
SEPM UC	–35%	–36%	100%	38%	–35%	18%	34%	32%	–15%	–11%	38%	34%	–42%
SEPM LOT	–1%	–31%	38%	100%	–34%	31%	28%	31%	–13%	–9%	65%	78%	7%
LOT MP	65%	55%	–35%	–34%	100%	–34%	–26%	–31%	23%	34%	–41%	–38%	22%
SEPM DEV	–8%	–16%	18%	31%	–34%	100%	16%	24%	–34%	4%	60%	40%	7%
WT	–7%	–27%	34%	28%	–26%	16%	100%	92%	–25%	6%	33%	51%	6%
DIAM	–8%	–21%	32%	31%	–31%	24%	92%	100%	–57%	8%	34%	52%	5%
DEN	17%	–8%	–15%	–13%	23%	–34%	–25%	–57%	100%	–11%	–18%	–17%	–3%
LLOT	32%	23%	–11%	–9%	34%	4%	6%	8%	–11%	100%	–1%	–3%	18%
WPN T100	4%	–46%	38%	65%	–41%	60%	33%	34%	–18%	–1%	100%	81%	0%
WPN T1000	18%	–49%	34%	78%	–38%	40%	51%	52%	–17%	–3%	81%	100%	12%
Q_n/Q_{max}	38%	37%	–42%	7%	22%	7%	6%	5%	–3%	18%	0%	12%	100%

APPENDIX F

Techniques for Developing Expenditure Profiles for SE/PM Development Costs

Cost estimators are sometimes asked for budgetary purposes to spread point estimates of weapons system development costs across a year-by-year spending profile. We reviewed our cost data for information that would be helpful in providing guidance for spreading point estimates of aircraft development programs. Please note that the following discussion relates to *expenditure* profiles for SE/PM costs. A second step is required to translate these expenditure profiles into budgeting profiles.[1]

For this analysis, we collected expenditure information on some of the more recent tactical aircraft development programs.[2] Unfortunately, we did not have CCDR data in regular increments for all or even most of the programs in our dataset. We decided to analyze the fighter aircraft development programs as a subset of our larger dataset

[1] Within government, there is a time delay from when budget authority is granted to when contracts are signed, funds are obligated, and finally funds are expended or there is an outlay of funds. DoD development programs are typically funded under Research, Development, Test, & Evaluation (RDT&E) appropriation, which allows funds appropriated in a single year to be expended over multiple years. The Under Secretary of Defense (Comptroller) typically estimates the outlay rates for each appropriation in budget estimates. Based on the National Defense Budget Estimate for FY 2004, the RDT&E defense-wide outlays are spread over four years in the following percentages: 46, 42, 10, and 2. Given the expenditure profile and the outlay rates, a system of linear equations can be used to determine the amounts that should be budgeted in each year. For more information, see Lee, Hogue, and Gallagher (1997).

[2] The names of the programs were withheld from this discussion to avoid supplying proprietary information. A supplemental, limited-distribution RAND report (TR-311-AF) supplies the name and data for each of the programs used for this analysis.

because we had incremental cost data on several older, full-scale development fighter programs as well as a few recent programs. This subset would allow us to hold constant the type of aircraft and nature of the development effort, while focusing on changes in duration and time.

Because the durations of the programs varied widely, we analyzed the rate of SE/PM expenditures using two approaches. The first approach uses a time scale based on the number of months from contract award. For this approach, we normalized the monthly time scale to a percentage of time, such that the duration starts with 0 percent of time completed at CA and ends with 100 percent of time completed at end of DT. The second approach uses milestone points as a means of gauging the amount of SE/PM costs expended at those points in the program schedule. Either approach can then be used against the total SE/PM development-cost point estimate to determine an expenditure profile.

Table F.1 shows the number of months to CDR, FF, and end of DT for the programs we analyzed. The first two programs showed a rather short duration of time to first flight. Upon investigation of these programs, we determined that the short schedule was due to work that was performed prior to the contract full-scale development program. We decided to not use the first two programs in our

Table F.1
Time to Development Milestones

Program	Months to CDR	Months to FF	Months to DT End	Months to CDR	Months to FF	Months to DT End
A	17	22	67	—	—	—
B	3	23	48	—	—	—
C	15	30	89	15	30	89
D	16	35	75	16	35	75
E	25	41	81	25	41	81
F	42	73	150	42	73	150
Average	20	37	85	25	45	99
Standard Deviation	13.0	18.9	34.8	12.5	19.4	34.6

analysis and concentrate on the other four programs that had schedules that seemed more consistent with recent aircraft development programs.

Using the first approach for determining the expenditure profile, we started the analysis by developing a Weibull[3] distribution for each program to model the actual expenditure profile against a normalized time scale. The plotted data for the four programs modeled by the Weibull distribution appear in Figure F.1. The modeled data are very close to the actual data points. Any missing data points, such as was the case for Programs E and F, were interpolated by the Weibull curve. The Weibull distribution essentially performed an optimization of distribution parameters to best fit the curve to the data points.

The next step was to determine an average Weibull curve that could be used to model SE/PM expenditures in a general fashion. The Weibull curves for each program were used to generate expenditures at common incremental points in the program's duration. The data at each increment were averaged across the four programs to develop an average SE/PM expenditure profile. By iteratively choosing shape and scale parameters for the average distribution and minimizing the error between a predicted expenditure curve and the average curve, we were able to generate average Weibull parameters. Table F.2 shows the Weibull shape and scale parameters generated for each program and for the average program.

A cost analyst can use these alpha and beta parameters to create a cumulative expenditure profile at desired increments of time in a spreadsheet program such as Microsoft Excel. Excel has a Weibull

---

[3] The Weibull distribution is similar to the Rayleigh distribution often used by cost estimators to estimate expenditure profiles of development programs. The Weibull is a more generalized form than the Rayleigh, with one additional adjustable parameter. The Weibull cumulative distribution function is: F(x,a,b) = 1 – e^ – (x/b)^a, where x is a point in the program's duration between 0 and 1, e is the constant 2.71828182845904 (the base of the natural logarithm), and a and b are adjustable parameters to the distribution that describe its scale and shape. The additional adjustable parameter in the Weibull allowed a slightly better fit of the historical data than the Rayleigh. Both functions fit the historical data closely and will provide a reasonably shaped budget profile for a development program.

Figure F.1
SE/PM Cost Spreads for Four Fighter Aircraft Development Programs, Normalized Durations

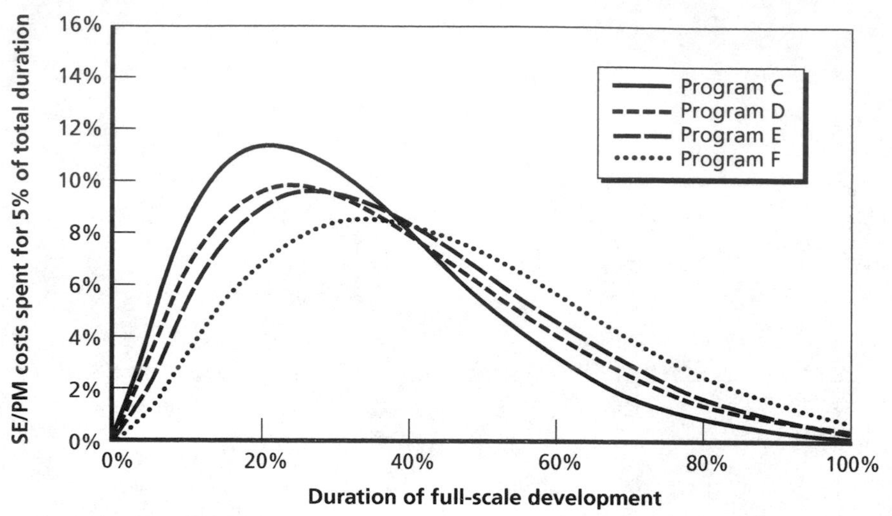

RAND *MG413-F.1*

Table F.2
Weibull Parameters for Modeling SE/PM Development Expenditures

Program	Alpha-Shape Parameter	Beta-Scale Parameter
C	1.6605	0.3418
D	1.6470	0.3952
E	1.7722	0.4127
F	1.9025	0.4839
Average program profile	1.7047	0.4079

function that requires an analyst to input the cumulative time elapsed and the alpha and beta parameters. The function returns the cumulative amount of funds spent at the specified time.

As stated above, with the second approach, we compared costs at commonly used milestone points. Table F.3 shows the cumulative percentage of elapsed program time at each milestone (with 0 percent being the contract award milestone and 100 percent being the end of DT milestone) and the cumulative percentage of SE/PM expended by CDR and FF on the four aircraft development programs. These data indicate that at CDR, which happens on average when 23 percent of the program's duration has elapsed, about one-third of the SE/PM funding is expended. At first flight, which occurs on average at 43 percent of the way through a program, about two-thirds of the SE/PM funds are spent. These findings indicate that the SE/PM funding is somewhat front-loaded against the schedule and that CDR and FF milestones can be used as good benchmarks for determining the amount of SE/PM expenditures expected at each milestone.

We used the same method of analysis for guided weapons development programs, but we had complete CCDR data for only one guided weapons full-scale development program. The expenditure profile for the guided weapons development program looked much like the profile for the aircraft programs. The Weibull shape and scale values were 1.920 and 0.4534, respectively. For this guided weapon program, the CDR and first guided launch dates occurred almost at the same time; CDR occurred 46 percent of the way through the program, and first guided launch occurred 48 percent of the way through the program. By the first guided launch, 62 percent of

Table F.3
Cumulative SE/PM Expenditures by Development Milestone

	CDR		First Flight	
Program	Elapsed Time	SE/PM Funds Expended	Elapsed Time	SE/PM Funds Expended
C	17%	27%	34%	62%
D	21%	30%	47%	73%
E	30%	44%	49%	75%
F	24%	30%	42%	64%
Average	23%	33%	43%	69%

SE/PM funds had been expended. Figure F.2 shows the plot of the Weibull expenditure profile for this program, normalizing the elapsed time to 0 percent at contract award and 100 percent at the end of DT.

In this appendix, we discussed methods for generating expenditure profiles for point estimates of SE/PM costs for aircraft and guided weapons development programs. An analyst may use either the Weibull function with the parameters listed above or the milestone approach. Average values can be used, or, if an analogous approach is deemed more appropriate, values from a specific program of interest can be used.

Figure F.2
SE/PM Cost Spreads for One Guided Weapons Development Program, Normalized Duration

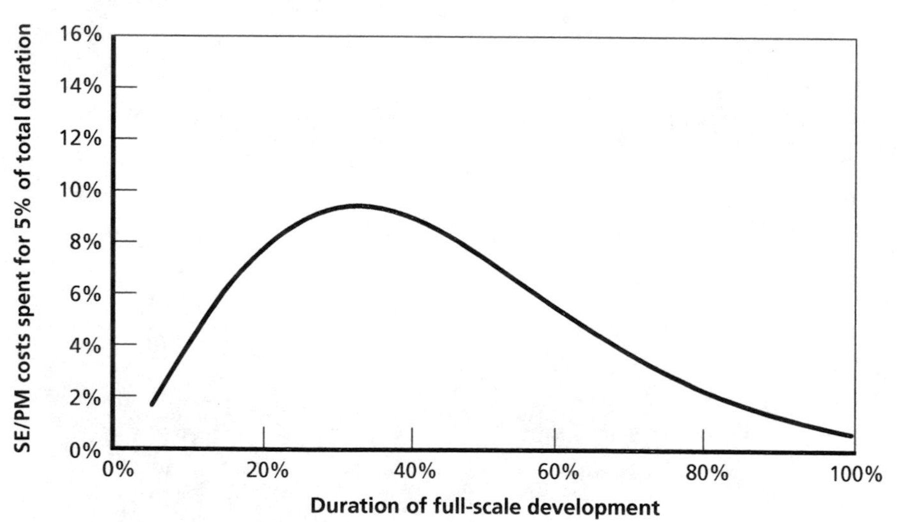

RAND *MG413-F.2*

Bibliography

Armament Product Group Manager, *Weapons File,* Eglin Air Force Base, Fla.: Air Armament Center, 1999.

Ball, Robert E., *The Fundamentals of Aircraft Combat Survivability Analysis and Design*, New York: American Institute of Aeronautics and Astronautics (AIAA) Education Series, 1985.

Book, Stephan A., and Philip H. Young, "The Trouble with R^2," *Journal of Parametrics,* forthcoming.

Chambers, George J., "The System Engineering Process: A Technical Bibliography," INCOSE (International Council on Systems Engineering), reprinted from *IEEE Transactions on Systems, Man, and Cybernetics,* Vol. SMC-16, No. 5, September/October 1986, pp. 712–722 (http://www.incose.incose.propagation.net/lib/sebib.html last accessed Aug 25, 2003).

Chromicz, J., *The Tri-Service Missiles and Munitions Automated Cost Database*, Arlington, Va.: Tecolote Research, Inc., 2001.

Cook, Cynthia R., and John C. Graser, *Military Airframe Acquisition Costs: The Effects of Lean Manufacturing,* Santa Monica, Calif.: RAND Corporation, MR-1325-AF, 2001.

Defense Acquisition University, *Defense Acquisition Acronyms and Terms,* Fort Belvoir, Va.: DAU, 2003.

——, *Defense Acquisition Guidebook,* Section 4.2.2, "Standards and Models," Fort Belvoir, Va.: DAU, December 2004a.

——, *Defense Acquisition Guidebook,* Section 4.3, "Systems Engineering Activities in the System Life Cycle," Fort Belvoir, Va.: DAU, December 2004b.

——, *Defense Acquisition Guidebook*, Section 10.3.3, "Industry Participation," Fort Belvoir, Va.: DAU, December 2004c.

——, *Program Managers Toolkit*, Washington D.C.: DAU, 2000.

Defense Systems Management College, "Concurrent/Systems Engineering Fact Sheet," *The Program Manager's Notebook,* Fort Belvoir, Va.: DSMC, July 1995.

DoD, Military Standard 499A, *Engineering Management*, 1974.

——, Military Standard 1521, *Technical Reviews and Audits for Systems, Equipments, and Computer Software,* 1995.

——, *DoD Handbook—Work Breakdown Structure*, MIL-HDBK-881, January 2, 1998 (dcarc.pae.osd.mil/881handbook/milhdbk881_cover_chap1.pdf; last accessed October 2005).

——, *The Defense Acquisition System,* DoD Directive 5000.1, May 12, 2003a.

——, *Operation of the Defense Acquisition System*, Department of Defense Directive 5000.2, May 12, 2003b.

DoD, Office of Undersecretary of Defense (Acquisition, Technology, and Logistics), *DoD Guide to Integrated Product and Process Development,* Version 1.0, February 1996 (http://www.acq.osd.mil/io/se/ippd/guide/ippd_concept.html; last accessed November 19, 2003).

——, *Rules of the Road, A Guide for Leading Successful Integrated Product Teams*, October 1999.

——, "Policy for Systems Engineering in DoD," memorandum, February 20, 2004a.

——, "Policy Addendum for Systems Engineering," memorandum, October 2004b.

DoD, Office of Undersecretary of Defense (Acquisition, Technology, and Logistics), Logistics Plans and Programs, *MILSPEC Reform: Final Report,* Fort Belvoir, Va.: April 2001.

DoD Directive 5000.4M, *Cost Analysis Guidance and Procedures,* December 11, 1992.

Fox, Bernard, Michael Boito, John C. Graser, and Obaid Younossi, *Test and Evaluation Trends and Costs for Aircraft and Guided Weapons,* Santa Monica, Calif.: RAND Corporation, MG-109-AF, 2004.

Freeman, J. V., *Naval Weapons Handbook,* China Lake, Calif.: Technical Information Division, Naval Air Warfare Center Weapons Division, 1999.

Harmon, Bruce R., J. Richard Nelson, Mitchell S. Robinson, Kathryn L. Wilson, and Steven R. Shyman, *Military Tactical Aircraft Development Costs, PA&E,* Alexandria, Va.: Institute for Defense Analyses, 1988.

Harmon, Bruce R., and Lisa M. Ward, *Methods for Assessing Acquisition Schedules of Air-Launched Missiles,* Alexandria, Va.: Institute for Defense Analyses, 1989.

Harmon, Bruce R., Lisa M. Ward, and Paul R. Palmer, *Assessing Acquisition Schedules for Tactical Aircraft,* Alexandria, VA: Institute for Defense Analyses, 1989.

Hess, Ronald Wayne, and H. P. Romanoff, *Aircraft Airframe Cost Estimating Relationships: Study Approach and Conclusions,* R-3255-AF, Santa Monica, Calif.: RAND Corporation, 1987.

Hoshour, Guy, et al., *SMC Systems Engineering Primer & Handbook,* Los Angeles, Calif.: U.S. Air Force, 2003.

Joint Chiefs of Staff Instruction, *Joint Capabilities Integration and Development System,* CJCSI 3170.01E, May 2005.

JSF Program Office, "JSF EMD Estimate Briefing," supplied to authors by JSF program office personnel, Arlington, Va., 1999.

Kain, Shawn M., *Alternative Methodologies for Estimating Systems Engineering/Program Management (SE/PM),* Wright-Patterson Air Force Base, Ohio: Aeronautical Systems Division, 1990.

Lee, David A., *The Cost Analyst's Companion,* McLean, Va.: The Logistics Management Institute, 1997.

Lee, David A., Michael R. Hogue, and Mark A. Gallagher, "Determining a Budget Profile from a R&D Cost Estimate," *Journal of Cost Analysis,* Nos. 29-41, Fall 1997.

Lorell, Mark A., and John C. Graser, *An Overview of Acquisition Reform Cost Savings Estimates,* Santa Monica, Calif.: RAND Corporation, MR-1329-AF, 2001.

Lowell, Stephen, "Beyond MilSpec Reform," presentation at the Senior Executive Service 2001 conference, Defense Standardization Program Office, 2001.

Luman, Ronald R., and Richard S. Scotti, "The System Architect Role in Acquisition Program Integrated Product Teams," *Acquisition Quarterly Review,* Fall 1996, pp. 83–96.

Martin, James N., "Overview of the EIA 632 Standard—Processes for Engineering a System," presentation to EIA 632 Working Group, September 1998.

McBride, Samuel W., *Missile and Munitions CER Development Study: Production Below-the-Line Cost Research*, Goleta, Calif., and Huntsville, Ala.: Tecolote Research, 2001.

Mendenhall, William, *Statistics for Management and Economics*, Sixth Edition, Boston: PWS-Kent, 1989.

Pagliano, Gary J., and Ronald O'Rourke, *Evolutionary Acquisition and Spiral Development in DoD Programs: Policy Issues for Congress,* Washington, D.C.: Congressional Research Service, 2003.

Perry, William, Secretary of Defense, "Specifications and Standards—A New Way of Doing Business," memorandum, June 29, 1994 (http://www.sae.org/standardsdev/military/milperry.htm; last accessed October 2005).

Pfleeger, Shari Lawrence, Felicia Wu, Rosalind Lewis, *Software Cost Estimation and Sizing Methods, Issues, and Guidelines*, Santa Monica, Calif.: RAND Corporation, MG-269, 2004.

Pingel, Walt, Global Hawk Program Office, "Estimating Implications of Spiral Acquisition: Global Hawk Case Study," DoD Cost Analysis Symposium briefing, Williamsburg, Va., January 29, 2003.

Przemieniecki, J. S., *Acquisition of Defense Systems,* Washington, D.C.: American Institute of Aeronautics and Astronautics, Inc., 1993.

Resetar, Susan A., J. Curt Rogers, and Ronald W. Hess, *Advanced Airframe Structural Materials: A Primer and Cost Estimating Methodology,* Santa Monica, Calif.: RAND Corporation, R-4016-AF, 1991.

SAS Institute, *SAS/STAT User's Guide*, Cary, N.C.: SAS Institute, Inc., 1999.

Sheard, Sarah A., and Jerome G. Lake, "Systems Engineering Standards and Models Compared," paper on Software Productivity Consortium, Systems and Software Consortium, Inc., Herndon, Va., 1997 (http://www.

software.org/pub/externalpapers/9804-2.html; last accessed August 25, 2003).

Wideman, Max R., *Wideman Comparative Glossary of Project Management Terms,* V2.1, May 2001 (http://www.pmforum.org/library/glossary/PGM_P07.htm; last accessed August 2004).

Younossi, Obaid, Mark V. Arena, Richard M. Moore, Mark A. Lorell, Joanna Mason, and John C. Graser, *Military Jet Engine Acquisition: Technology Basics and Cost-Estimating Methodology*, Santa Monica, Calif.: RAND Corporation, MR-1596-AF, 2002.

Younossi, Obaid, Michael Kennedy, and John C. Graser, *Military Airframe Costs: The Effects of Advanced Materials and Manufacturing Processes,* Santa Monica, Calif.: RAND Corporation, MR-1370-AF, 2001.

Younossi, Obaid, David E. Stem, Mark A. Lorell, and Frances M. Lussier, *Lessons Learned from the F/A-22 and F/A-18E/F Development Programs,* Santa Monica, Calif.: RAND Corporation, MG-276, 2005.